Developments in Turbomachinery Flow

Related titles

Schobeiri, *Turbomachinery Flow Physics and Dynamic Performance,* Second edition, Springer, 2012, 9783540223689

Lakshminarayana, *Fluid Dynamics and Heat Transfer of Turbomachinery,* Wiley-Interscience, 1995, 9780471855460

Woodhead Publishing in Mechanical Engineering

Developments in Turbomachinery Flow

Forward Curved Centrifugal Fans

Nader Montazerin
Department of Mechanical Engineering
Amirkabir University of Technology

Ghasem Akbari
Industrial and Mechanical Engineering Faculty
Islamic Azad University, Qazvin Branch

Mostafa Mahmoodi
Department of Mechanical Engineering
Amirkabir University of Technology

ELSEVIER

AMSTERDAM • BOSTON • CAMBRIDGE • HEIDELBERG
LONDON • NEW YORK • OXFORD • PARIS • SAN DIEGO
SAN FRANCISCO • SINGAPORE • SYDNEY • TOKYO
Woodhead Publishing is an imprint of Elsevier

WP
WOODHEAD PUBLISHING

Woodhead Publishing is an imprint of Elsevier
80 High Street, Sawston, Cambridge, CB22 3HJ, UK
225 Wyman Street, Waltham, MA 02451, USA
Langford Lane, Kidlington, OX5 1GB, UK

British Library Cataloguing-in-Publication Data
A catalogue record for this book is available from the British Library

Library of Congress Control Number: 2015932043

ISBN 978-1-78242-192-4 (print)
ISBN 978-1-78242-193-1 (online)

For information on all Woodhead Publishing publications visit our website at http://store.elsevier.com/

Printed and bound in the United Kingdom

To:

Roshanak, Atekeh, and Zeinab

Contents

List of figures

List of tables

About the authors

Nader Montazerin is professor of mechanical engineering at Amirkabir University of Technology. He received his BS from Sharif University of Technology, Tehran, Iran (1980) and MS and PhD from Cranfield University of Technology, Bedford, England (1982 and 1985). He immediately started lecturing and research at Amirkabir University of Technology, where his interests are centred on experimental fluid mechanics, turbomachines, compressible flow and applied energy. He has published five books and many journal articles. His industrial activities are on gas turbines, national gas networks and energy efficiency.

Ghasem Akbari is assistant professor of mechanical engineering at Islamic Azad University, Qazvin branch, Industrial and Mechanical Engineering Faculty. He has held this position since 2014. He received his PhD from Amirkabir University of Technology, Tehran, in 2014. He has taught several courses at Islamic Azad University including fluid mechanics, heat transfer, thermodynamics and internal combustion engines. He has served on several provincial/academic projects, including the identification of gas consumption bottlenecks as well as its long-term prediction. He is also interested in experimental and theoretical fluid mechanics and turbulence, particularly in turbomachinery applications. He has published several prominent journal articles on the results of research during his PhD period.

Mostafa Mahmoodi is a PhD candidate in mechanical engineering at Amirkabir University of Technology. He received his BS in mechanical engineering – thermo fluids from Yazd University, Yazd, Iran, in 2007. He received his MS in mechanical engineering – energy conversion from University of Kashan in 2011. His PhD thesis is on numerical and experimental study of aeroacoustic behaviour of squirrel cage fans.

Preface

Squirrel-cage fans are strange. Precious electricity produced by gas and steam turbines is lost to the very inefficient forward-curved fans of the heating and ventilating industry. The research studies on gas and steam turbines are expensive and very competitive worldwide. On the contrary, forward-curved squirrel-cage fans are cheap and simple to build which makes any application of research findings very simple.

The study of the complicated flow field requires advanced measuring techniques such as laser Doppler anemometry. The combination of building different fans, doing performance tests and examination of the flow field with laser anemometry was an attractive research field that was carried out for many years by the team that had written the present book.

Early in this research, modifications to the inlet were proved to be productive. The improvements were far more significant than what was already reported in the literature for other modifications. Later, the present team started with the rotor and the idea of a half-cone rotor and a half-cone with a lean angle that proved to have merit in many cases. Tests on the volute geometry indicated that it was not the shape of the volute cross-section that is important, but the spread angle. Larger spread angle could result in larger flow rates and could override all optimizations to the inlet! Still an optimum spread angle could be defined that gives an aerodynamic flow for a limited space.

It was soon known that optimizations without modelling are very expensive and time-consuming. Simulation of the flow in this fan is not an easy task. Three-dimensional flow, separated flow after the inlet, large number of blades and separated three-dimensional boundary layer inside the blades were all the most important complications in any simulations. Although modelling is not accurate, it was proved to be successful in determining the trends. The modellings were managed to direct research for correct experimental setups and minimum costs.

With the gained experience on fans and laser anemometry, it was natural to move into stereoscopic particle image velocimetry. Squirrel-cage fans are ideal instruments for sample flow measurements in centrifugal turbomachines. They give flat sidewalls and easy access to measuring plane. They rotate slowly and have many features like separations and jet/wake flows that have baffled research studies for a long time.

It was a fortunate occasion that Chandos Publishing invited us to write a book on such fans at this stage. This opportunity provided us with the best chance to review what was achieved during the past twenty-four years. The first four chapters of the book contain our studies on fan performance, fluid flow and fan optimization. Chapter 5 deals with aeroacoustic and sound generation behaviour of the forward-curved centrifugal fans. Chapter 6 investigates the contribution of jet–wake–volute interactions to flow characteristics and turbulence models which are more advanced, and we expect it to find serious application when the computing capacity for simulating centrifugal turbomachines increases.

We are grateful to acknowledge prior publication of a part of our research in the *Journal of Power and Energy, Journal of Turbulence* and *International Journal of Heat and Mass Transfer*. Detailed references are given at the end of each chapter.

I (the first author) have enjoyed the work of two excellent PhD students in the past few years who kindly accepted to be my coauthors in this book. Dr Akbari defended his PhD in 2014 and Mr Mahmoodi shall complete his PhD in a couple of years. There are still many of my old students who contributed to the subject presented in this book and I just name them here with much respect:

Mr Matin Hosseini; Mr Mohammad Reza Shetab Booshehri; Mr Mohammad Reza Najjari; Mr Seyyed Alireza Hosseini; Dr Reza Sepahi Samian; Mr Hossein Akhavan; Mr Soroosh Shahabadi; Mr Masood Nikkhoo; Late Mr Morteza Akbarizadeh; Mr Javad Alinejad; Mr Sina Samarbakhsh; Mr Seyyed Mohammad RezaeiNiya; Mr Arash Heshmat Dehkordi; Mr Saeed Zeinali; Dr Ebrahim Damangir; Mr Mohammad Hassan Hessami Azizi; Mr Mojtaba Gholamian; Mr Mehdi Askari Shahi; Mr Ahmad Kazemi Fard; Mr Ali Toorani; Mr Mohammad Jafar Mahmoodi; Mr Hamid Reza Mirzaei; Dr Seyyed Saeed Mirian; and Dr Ehsan Tooyserkani.

The work is also incomplete if we do not mention other researchers who have done parallel works on squirrel-cage fans round the globe. As far as we know, the only team who has produced continuous research in this field is from Oviedo University in Spain. Their articles have been of much interest and inspiration to us. Other researchers have not been consistent and only produced occasional results. We are indebted to all the inspirations we received from these groups and have tried to cite their works at the end of each relevant chapter.

Nader Montazerin
Tehran
March 2015

General introduction of forward-curved squirrel-cage fan

1

1.1 Introduction

Forward-curved squirrel-cage fans are widely used for residential and automobile air conditioning and also for cooling of computers and electronic appliances. High volumetric flow rate, low noise and simple construction are basic characteristics of such fans. They are easy to manufacture, operate quietly and pump a relatively large volume of air. The squirrel-cage fans are slower than other fans for the same flow rate, or are smaller for the same head. This feature and low manufacturing cost make them very popular in the comfort industry and household appliances.

Such merits are achieved at the cost of low efficiency, which is related to the poor aerodynamics of squirrel-cage fans. These fans are usually designed for a lower first cost and, therefore, the efficiency is not higher than 50–60%. Although individual fan power consumption is low, since there are a large number of such appliances in use, the overall energy consumption becomes significant. Even small improvements in their performance, when multiplied by their widespread application, are important for energy saving.

The squirrel-cage fan is a strange piece of machinery. It is widely used but far less studied. It works with considerable volume of separated flow inside it. In fact, as shall be shown in this book, in some cases larger separation is more beneficial to fan characteristics. It is found that the squirrel-cage fans are excellent devices for examination of the flow phenomena of turbomachines.

Evaporative coolers were the initial drive for the authors to pay attention to such fans. A geometrical optimization was performed as the initial step of fan operation enhancement. It was discovered rapidly that interrelation between different dimensions makes the fan very unpredictable. Three-dimensional flow and separation are obstacles to accurate modelling. A combination of modelling and experiment was then followed.

Fans are judged through their characteristic curves. This is the criterion for the customer to select a fan. Every modification in fans does not necessarily affect their characteristics. An intermediate step to improve centrifugal fans is to reach uniform flow through the blades and volute and to reduce turbulence intensities. Other industrial criteria are also cited that include compactness and lower sound-pressure levels.

Available research on the squirrel-cage fan includes attempts to understand the complicated flow structure within the fan, and in this way covers the following four different viewpoints:

1. Parametric performance studies on the classic fan configuration for different geometries. In these studies the rotor diameter is set as a base for fan design and performance is optimized with variations in rotor or volute parameters.

Developments in Turbomachinery Flow.

2. The second group of research tries to understand flow fields inside the fan and relates the results to observations on performance. This includes velocity measurements with Pitot tube, hot-wire and laser Doppler anemometers. Such measurements are only possible for limited points within the fan and for certain points on the performance curves. The results are successful in showing the flow separation after the inlet, jets and wakes after the blades, and the nonuniform flow through the volute.
3. The third group includes attempts to increase efficiency by modifying the common configuration, for example, adding suction just after the inlet for a smaller separation as well as using an outward or annular inlet. Complication in such manufacturing is an important feature that should be considered in each new configuration.
4. Although small fan noise is an inherent advantage of such fans, the fourth group of studies on noise has suggestions for geometric parameters that would result in a quieter fan. These relate fan noise to flow field and geometry through simultaneous noise and turbulence measurements.

1.2 Fan geometry

A schematic of the forward-centrifugal fan is given in Figure 1.1. It is a long time that people do not attempt to model forward-blade fan performance with the Euler energy equation. Flow-field studies in this book and from other researchers have clarified that the flow is not uniform around the impeller or across its width. Large portions of impeller width are occupied with reversing flow or zero radial velocity. Chapter 4 of this book shows extensive variation in slip factor over the impeller exit surface that means the flow does not follow the blade passages. Presenting an averaged Euler equation would just miss flow complexities inside the fan and gives an average which is neither applicable to design nor helpful for research. Unlike other turbomachines, separations inside the blade passages prohibit the use of the Euler energy equation to model the flow field for a corresponding head or power modelling. Results are available for modifications in rotor geometry such as aerodynamic blade profiles or a removal of the shroud but since the rotor is the most expensive part of the fan, it is very important that modifications should not result in added fan price.

There are suggested optimum values for squirrel-cage rotor dimensions such as diameter to width ratio, inner and outer diameters and blade inlet or outlet angles. Typical impeller diameters range from 4 to 80 cm. The impellers have 30–40 short-chord blades that are manufactured from thin sheet plates. Inner to outer diameter is between 0.7 and 0.8, and width to diameter ratio is about 0.35–0.4. Blade inlet and outlet angles are about 90° and 175°, respectively, and the optimum number of blades is about 40. It is worth noting that unlike many turbomachines, these are not parameters that change from machine to machine based on its load or other specifications. Tests have shown that the above-mentioned range of geometrical parameters produce higher nondimensional performances such as pressure coefficient, flow coefficient or efficiency. The only condition is that machine Reynolds number should be larger than 4×10^5 in order to avoid a laminar flow in the cascade. This condition is usually met unless the rotor is small or it rotates slowly.

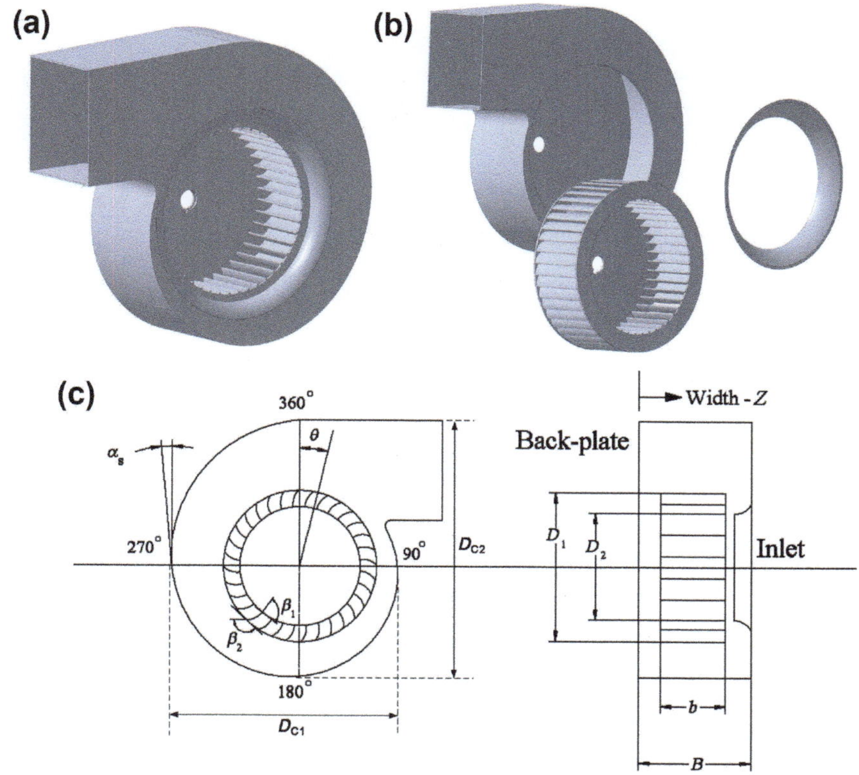

Figure 1.1 Geometry of a squirrel-cage fan: (a) three-dimensional geometry of the assembled fan; (b) three-dimensional geometry of the main parts of the fan; (c) two-dimensional views of the fan including the assigned notations in this book.

A larger volute spread angle results in larger flow rates when all other parameters like rotor size and its rotational speed are held constant. It will be shown in Chapter 4 that this is not due to a more aerodynamic flow inside the fan. The fan is not sensitive to the width ratio between the rotor and the volute if this ratio is near 1. Wider volutes reduce performance and initiate instabilities. Such instabilities are not still surveyed or documented but moving the rotor away from the inlet could initiate them even at high flow rates.

The Cordier diagram compares performance and size of different turbomachines (Figure 1.2). This diagram links the optimum operating conditions (the volumetric flow rate, Q, and the specific head, $\Delta p_t/\rho$, at the optimum efficiency, η) with the diameter, D_2, and rotational speed, n, using nondimensional velocity, σ, and nondimensional size, δ, that are defined as

$$\sigma = \frac{\varphi^{\frac{1}{2}}}{\psi_t^{\frac{3}{4}}} = \frac{\sqrt{\pi}2^{\frac{1}{4}}n\sqrt{Q}}{\left(\Delta p_t/\rho\right)^{\frac{3}{4}}} \tag{1.1}$$

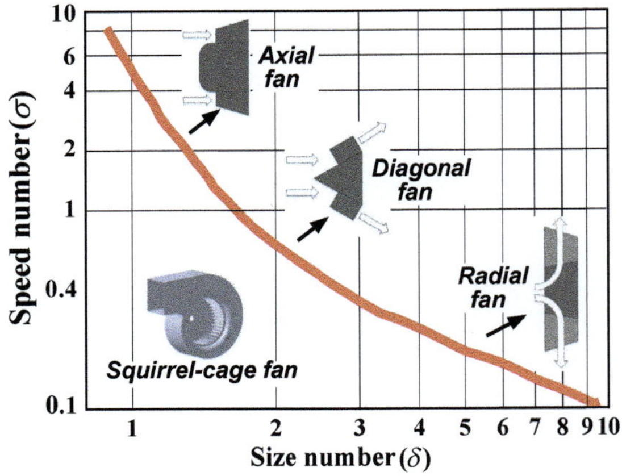

Figure 1.2 Representation of different types of turbomachines on the Cordier diagram.

$$\delta = \frac{\psi_t^{\frac{1}{4}}}{\varphi^{\frac{1}{2}}} = \frac{\sqrt{\pi} D_2 (\Delta p_t/\rho)^{\frac{1}{4}}}{2^{\frac{3}{4}} \sqrt{Q}} \tag{1.2}$$

where φ and ψ_t are flow coefficient (nondimensional flow rate) and total head coefficient (nondimensional total pressure), respectively, and are defined later. Flow direction in the meridional plane locates different fans in the Cordier diagram. Axial fans are in the higher-energy side with smaller dimensions and higher rotating speeds. Radial fans rotate slower but have larger diameters. There is no boundary between fully axial or fully radial categories. Squirrel-cage fans rotate faster for similar load as radial fans but have a similar size as axial machines.

1.3 Flow field

The research on squirrel-cage fans includes attempts to understand the flow field inside the fan. Three-dimensional interactions among vortical flow structures significantly contribute to fluid mechanics of such machines. The wetted area in these machines is relatively large and friction retards the flow considerably. Like any other radial turbomachine, flow experiences intensive swirl and generates a complex secondary flow inside the turbomachine. Reduction of efficiency is a consequence of such complicated flow and design of a high-efficiency squirrel-cage fan is a difficult task.

Figure 1.3 shows that the flow enters the fan axially from the shroud side and gradually turns into the radial direction. The bell-mouth inlet directs the flow to the rotor where it is expected to turn up to 90°. Usually this does not happen for about one-third of the rotor width and, therefore, the flow inlet angle in the meridional plane is not perpendicular. High incidence, mismatch between inlet and impeller and nonuniform flow at impeller inlet result in extensive unsteady separation over the blade suction side and inside the volute. Even the best available configuration gives poor flow guidance

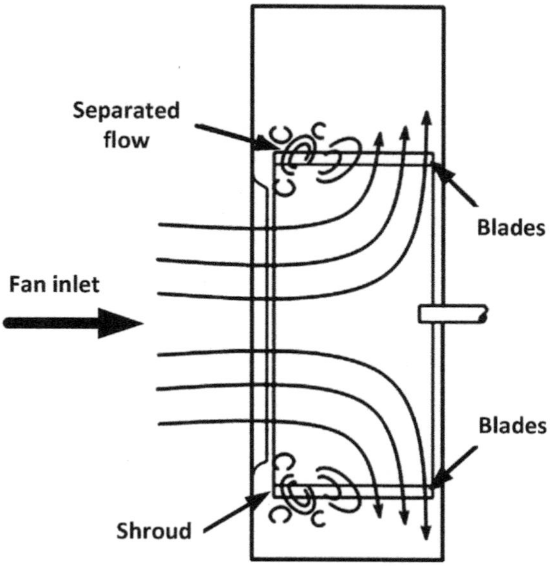

Figure 1.3 Interactive region of flow behind the inlet.

and inefficient energy exchange inside the fan. From the shroud to one-third of the impeller width, there are regions of two- and three-dimensional separation with no through-flow. Two-dimensional velocity measurements inside the volute of different squirrel-cage fans confirm such behaviour and show the separated flow behind the inlet and nonuniform velocity profile along the blade or around the rotor. This characteristic is the prime cause of the low efficiency in these fans.

At the other two-thirds of the rotor width, through-flow happens and a jet—wake pattern occurs downstream of the impeller. Even at the best efficiency point, the velocity components and their fluctuations (inside and outside the impeller) depend on axial and circumferential locations and the recirculation from the volute towards the impeller (around the cut-off region).

The important case in forward-curved squirrel-cage fans is that any modification in its aerodynamics should not add to the manufacturing cost. There is still little knowledge about the flow field inside such fans at different geometries and various flow conditions. The nonlinear characteristic of the problem makes a strict dependency between flow solution and the initial conditions. For such a sensitive problem, the optimization of geometry is also quite dependent on various geometrical parameters. In other words, if there is an optimum for one geometry, it is no longer efficient if one basic dimension changes. This is discussed in more detail throughout the book and, for example, it is shown that inlet optimization becomes meaningless for larger volute spread angles.

The results presented in this book are extracted from four different approaches:

1. Fan performance and noise measurements: This is what the individual customers or the system designers require.

2. Simulation of the flow field inside the fan: The capability of this approach in exact simulation of turbomachine complex flow within a reasonable CPU time is arguable. Nevertheless, it has proved as a successful approach in predicting trends which in turn has reduced the test costs.
3. Laser Doppler anemometry (LDA): This has given point-wise/temporal velocity variations and velocity profiles just outside the rotor. This important station shows all inlet and rotor effects and would also be considered as the initial condition for the volute flow.
4. Particle image velocimetry (PIV): This is an up-to-date research-oriented method that gives further insight into flow complications in this fan. It is implemented on a classic geometry of squirrel-cage fan to investigate the complicated jet–wake flow out of the rotor, which could have implications for other centrifugal turbomachines.

These techniques are introduced in the following sections, and the results are analysed in the rest of the book.

1.4 Fan performance and noise measurements

Fan performance or characteristic fan curves specify the elevation in fluid head (pressure) and efficiency. These curves are used to compare various designs and help selection of the appropriate design for the application. Precise test procedures are developed by international standards such as BS848, ISO5801 and AMCA210. The fan performance procedure ISO5801 is selected and introduced in the following sections and applied throughout this book.

1.4.1 Standard installation types

Four types of installation can be used for fan performance tests:

- Type A: Free inlet and free outlet
- Type B: Free inlet and ducted outlet
- Type C: Ducted inlet and free outlet
- Type D: Ducted inlet and ducted outlet

The free inlet and ducted outlet (type B) is the most practical configuration in ventilation applications. Consequently, the performance tests given in this book are carried out on the basis of this configuration. It uses an outlet duct with a straightener to dampen the swirl in the fan outlet. The pressure is generally measured in the outlet duct after the straightener blades. The fan should be tested with no auxiliary device at the inlet.

1.4.2 Fan characteristic curves

Fan efficiency and elevations in static and total pressure are normally plotted versus fan flow rate. Pressure and flow rate are normalized into pressure coefficient (ψ) and flow coefficient (φ), respectively, in order to generalize the characteristic curves:

$$\varphi = \frac{4Q}{\pi^2 n D^3} \tag{1.3}$$

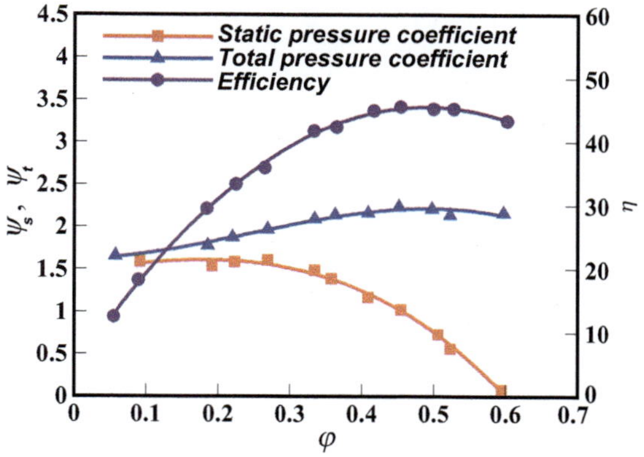

Figure 1.4 Characteristic curves of a forward-curved centrifugal fan with cylindrical rotor.

$$\psi = \frac{2P}{\rho \pi^2 n^2 D^2} \tag{1.4}$$

where n is the fan rotational speed (in rpm) and D is the channel diameter at the measurement section. The fan efficiency is defined as the ratio between power delivered to the airflow and power consumed by the fan:

$$\eta = \frac{k_p P_t Q}{P_r} \tag{1.5}$$

where k_p is the compressibility coefficient, P_t is total pressure and P_r is the motor power.

The optimal operating condition of a fan can be determined after plotting fan efficiency versus flow coefficient. Figure 1.4 presents typical characteristic curves of a forward-curved centrifugal fan with cylindrical rotor. The scattered data are from measurements and the continuous lines are just curve fits.

A few comments on possible configurations of characteristics are worth noting:

1. Some fans show unstable or bistable performance at smaller flow rates. Forward-curved fans, such as those studied in this book, may show such behaviour at higher flow rates. There is no criterion for avoiding such instabilities and only performance tests can show them.
2. There are points of maximum head or efficiency for some fans, but there are others with continuously increasing or decreasing characteristic curves.
3. Fans are usually selected to work in their right-half of characteristic curves. Therefore, a better performance in the left half is not advantageous and intensifies instabilities.

There is no design point to specify. Fan and system curves are usually matched to determine the operating point or operation range. This makes the flatter characteristic curves more desirable since there shall be less head or efficiency change in different operating points.

1.4.3 Standard installation for fan sound measurement

The proposed procedure by ISO5136 is used to study sound generation in fans. The setup in this standard is similar to the following installations in ISO5801:

- Type B: Free inlet and ducted outlet
- Type C: Ducted inlet and free outlet
- Type D: Ducted inlet and ducted outlet

In the last installation type, the microphone can be located in inlet or outlet ducts. Therefore, there are four different arrangements for noise experiments. The test rig on the basis of type B is the most suitable configuration for the forward-curved centrifugal fans. A schematic of this installation is shown in Figure 1.5. The fan is connected to the test duct using a transition duct. Some sound reflections occur at the end of the test duct, which are removed using an anechoic termination. A throttling cone adjusts the flow rate.

After recording the acoustic pressure signals by the microphone, the sound-pressure level (SPL) is calculated as

$$\text{SPL} = 10 \log_{10}\left(\frac{p_{rms}^2}{p_0^2}\right) \tag{1.6}$$

where p_{rms} is the root mean square value of sound-pressure signals and p_0 is the reference value equal to 20 μPa. The frequency spectrum of SPL is obtained by applying fast Fourier transform (FFT) on the results of Eqn (1.6).

1.5 Fluid-flow simulation in centrifugal fans

Experimental techniques are valuable tools for extracting the realistic flow-field data. However, they are limited to a particular region or even point, and their temporal data sampling is not necessarily as fine as desired. A general evaluation of the three-dimensional flow structures within the entire turbomachine domain would be possible by recent developments in computational fluid dynamics (CFD). The unsteady fluctuations of the turbomachinery fluid flow have significant drawbacks such as vibration of the casing and blades, deterioration of performance, generation of stall cells within the blades and aerodynamic noise. Simulation of the turbomachinery flow field helps in

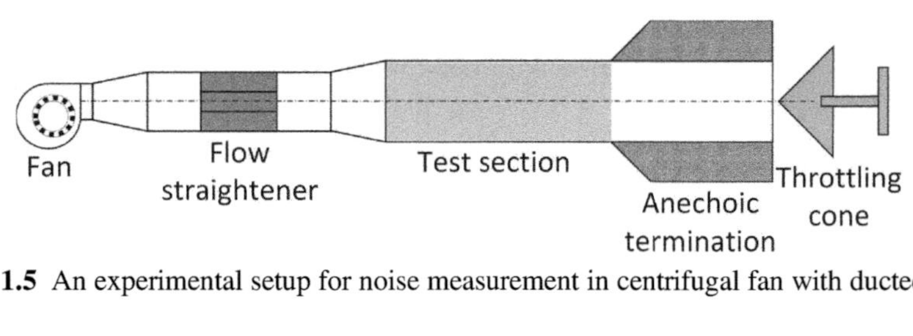

Figure 1.5 An experimental setup for noise measurement in centrifugal fan with ducted outlet.

understanding these phenomena and is a step towards removing or reducing such undesirable effects. However, this is a quite difficult task for a complicated geometry with numerous stationary/moving blades and extensive separated flow regions. Practical approaches for numerical simulation of such complicated flow field are discussed in the following.

1.5.1 Turbulence modelling

Direct numerical simulation (DNS) of the basic Navier–Stokes equation without any assumption or simplification is the ideal approach to obtain the actual flow field. However, the implemented numerical grid in the DNS should be sufficiently fine to capture maximum turbulent kinetic energy from small-scale flow structures. The count of computational elements in the simulation domain of high Reynolds number flows is roughly estimated to be proportional to $Re^{9/4}$. In the case of complicated geometries such as forward-curved squirrel-cage fans even smaller grid dimensions are required. Implementation of such grid resolution is not possible at present due to limitations of current computational resources. Simulation of turbulent flows on the basis of various decompositions and their respective turbulence models is the practical alternative for turbomachinery applications. Reynolds-averaged Navier–Stokes (RANS) and large eddy simulation (LES) are two popular model-based approaches for simulating turbulent flows. The choice of turbulence model has a strong influence on prediction of flow field in any turbomachine zone in the presence of separating/reverse-flow structures. The numerical prediction of aerodynamic performance of fans can be appropriately done with the steady RANS approach. The unsteady Reynolds-averaged Navier–Stokes (URANS) approach is a better choice for highly unsteady flow of the squirrel-cage fan. However, this approach is very sensitive to the initial and boundary conditions and should be done with special care.

The RANS approach models the Reynolds stress tensor by substituting the entire turbulence energy cascade with an appropriate model. The major complication is the modelling of large-scale turbulent-flow structures that strictly depend on device geometry and have no universal characteristics. This characteristic reduces the accuracy of the RANS approach in predicting Reynolds stresses. The LES approach can fulfil this drawback for the price of more computational costs. The LES directly simulates directional large-scale flow structures by solving the filtered Navier–Stokes equations and accounts for the contribution of small-scale (sub-grid-scale (SGS)) structures via modelling the SGS stress tensor. Since the small-scale flow structures are less affected by geometry and have more universal characteristics, these structures have a better potential for accurate modelling. This is the main benefit of the LES approach that causes more accurate modelling of the turbulence stress tensor as compared with the RANS approach.

Despite the progress in flow simulation, the LES approach is still not able to accurately predict the SGS stress in complicated anisotropic flow of the turbomachines and requires further modifications. The turbomachinery flow field is significantly unsteady and covers a diverse range of temporal time scales (ranging from a fraction of blade passing period to the Kolmogorov time scale). The spatial variation of turbomachinery flow structures is also very extensive and ranges from a fraction of blade-to-blade distance to the Kolmogorov length scale. Presence of three-dimensional anisotropic flow structures and existence of nonregular turbulent energy transfer mechanisms in the turbomachinery flow causes significant deviation between the modelled and real SGS stress tensor. According to such features as well as enormous computational cost of LES in comparison with RANS, different versions of the latter approach are more popular in simulation of squirrel-cage fans. In the following, the RANS and LES approaches are introduced, and their common models are evaluated.

1.5.1.1 RANS simulation

The accuracy of various RANS models in the squirrel-cage fan is evaluated by implementing $k-\varepsilon$ standard, $k-\varepsilon$ RNG and $k-\omega$ models and comparing the results with experimental performance data as shown in Figure 1.6. Performance parameters are global, and the ability to model them is a minimum requirement for any simulation software. It should be noted that the experimental results are obtained in a setup according to the ISO5108 and a similar computation domain is designed for numerical simulations. The results of the $k-\varepsilon$ RNG model are in reasonable agreement with the experimental data. However, at very low flow rates, the results of all RANS models are different from experimental results. Strong separations, reverse-flow zones and stall cells are challenging characteristics of flow field at very low flow rates that cannot be modelled precisely by any of the RANS models. The $k-\varepsilon$ RNG model is briefly introduced in the following.

The continuity and momentum equations in multi-reference frame are expressed as follows:

$$\nabla \cdot u_R = 0 \tag{1.7}$$

$$\nabla \cdot (u_R \otimes u_R) + 2\Omega \times u_R + \Omega \times \Omega \times R = -\nabla\left(\frac{p}{\rho}\right) + \nu\nabla \cdot \nabla u_R \tag{1.8}$$

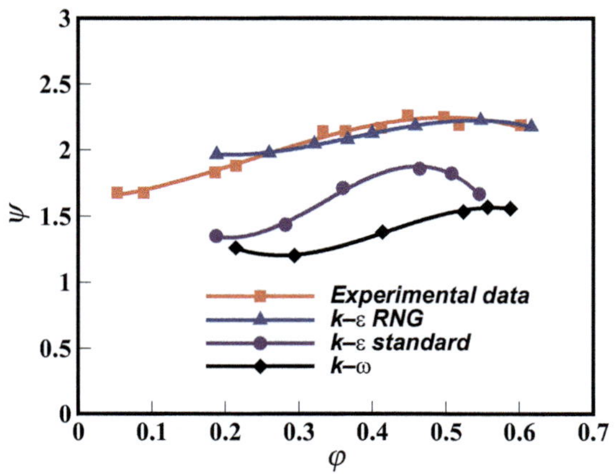

Figure 1.6 Comparison of performance of a squirrel-cage fan obtained by experiment and different Reynolds-averaged Navier–Stokes models.

where u_R is the flow velocity in a frame that rotates at Ω speed, R is the position vector, p is fluid pressure, and ρ and ν are density and kinematic viscosity of fluid, respectively. Two equations should be solved in the $k\!-\!\varepsilon$ RNG model in order to obtain turbulent kinetic energy (k) and viscous dissipation (ε):

$$\frac{\partial}{\partial x_i}(\rho u_i k) = \frac{\partial}{\partial x_i}\left(\alpha_k \mu_{eff} \frac{\partial k}{\partial x_i}\right) + G_k - \rho\varepsilon \tag{1.9}$$

$$\frac{\partial}{\partial x_i}(\rho u_i \varepsilon) = \frac{\partial}{\partial x_i}\left(\alpha_\varepsilon \mu_{eff} \frac{\partial \varepsilon}{\partial x_i}\right) + C_{1\varepsilon}\frac{\varepsilon}{k}G_k - C_{2\varepsilon}\rho\frac{\varepsilon^2}{k} \tag{1.10}$$

G_k represents the generation of turbulent kinetic energy due to mean velocity gradients:

$$G_k = \mu_t\left(\frac{\partial u_j}{\partial x_i} + \frac{\partial u_i}{\partial x_j}\right)\frac{\partial u_j}{\partial x_i} \tag{1.11}$$

The coefficients $C_{1\varepsilon}$ and $C_{2\varepsilon}$ are model parameters that are empirically derived as $C_{1\varepsilon} = 1.42$ and $C_{2\varepsilon} = 1.68$. The quantities α_k and α_ε are the inverse effective Prandtl numbers for k and ε, respectively, with the values $\alpha_k = \alpha_\varepsilon \approx 1.4$. Turbulent viscosity ($\mu_{eff}$) is calculated from

$$\mathrm{d}\left(\frac{\rho^2 k}{\sqrt{\varepsilon\mu}}\right) = 1.72\frac{\widehat{\nu}}{\sqrt{\widehat{\nu}^3 - 1 + C_\nu}}\mathrm{d}\widehat{\nu} \tag{1.12}$$

where $\widehat{\nu} = \mu_{eff}/\mu$ and $C_\nu \approx 100$. In the high Reynolds number limit, turbulent viscosity is given by

$$\mu_t = -\rho C_\mu \frac{\varepsilon^2}{k} \tag{1.13}$$

which is the same as the standard $k\!-\!\varepsilon$ approach with empirically determined constant $C_\mu \approx 0.09$.

1.5.1.2 Large eddy simulation

In LES, a filter is applied on the velocity field, $\mathbf{U}(\mathbf{x},t)$, and decomposes it into resolved and SGSs:

$$\overline{\mathbf{U}}(\mathbf{x},t) = \int_D G_\Delta(\mathbf{r},\mathbf{x})\mathbf{U}(\mathbf{x}-\mathbf{r},t)d\mathbf{r} \tag{1.14}$$

where D is the computational domain, G_Δ is the kernel of the filter with a characteristic scale Δ and the over-bar indicates filtering operation. The filtered continuity and Navier–Stokes equations for incompressible flow are

$$\frac{\partial \overline{u_i}}{\partial x_i} = 0 \tag{1.15}$$

$$\frac{\partial \overline{u_i}}{\partial t} + \overline{u_j}\,\frac{\partial \overline{u_i}}{\partial x_j} = -\frac{\partial}{\partial x_j}\left(\frac{\overline{p}}{\rho}\delta_{ij} + \tau_{ij}\right) + \nu\frac{\partial^2 \overline{u_i}}{\partial x_j^2} \tag{1.16}$$

where τ_{ij} is the SGS stress that is defined as

$$\tau_{ij} = \overline{u_i\,u_j} - \overline{u_i}\,\overline{u_j} \tag{1.17}$$

The momentum equations must be closed by expressing the SGS stress tensor as a function of resolved velocity components. The filtered Navier–Stokes equations are directly simulated to obtain resolved velocity field while SGSs are modelled.

The Smagorinsky model is one of the most popular SGS models that directly relates the deviatoric part of the SGS stress tensor ($\tau_{ij}^D = \tau_{ij} - \tau_{kk}\delta_{ij}/3$) to the resolved strain rate tensor ($\overline{S_{ij}}$):

$$\tau_{ij}^{D,\,Smag} = -2\nu_t^{Smag}\,\overline{S_{ij}} \tag{1.18}$$

where the Smagorinsky eddy viscosity (ν_t^{Smag}) and $\overline{S_{ij}}$ are defined as

$$\nu_t^{Smag} = \left(C_{Smag}\Delta\right)^2\left|\overline{S}\right| \tag{1.19}$$

$$\overline{S_{ij}} = \frac{1}{2}\left(\frac{\partial \overline{u_i}}{\partial x_j} + \frac{\partial \overline{u_j}}{\partial x_i}\right) \tag{1.20}$$

$\left|\overline{S}\right| = \sqrt{2\overline{S_{ij}}\,\overline{S_{ij}}}$ and C_{Smag} are the characteristic filtered strain rate (also known as strain magnitude) and the Smagorinsky coefficient, respectively.

Although the Smagorinsky model is frequently used in simulations, this model has drawbacks:

- An eddy-viscosity model such as the Smagorinsky model is completely dissipative and cannot predict energy transfer from SGS structures towards larger resolved scales (back-scatter of energy). Since the turbomachinery flows are significantly attributed by backscatter of energy, it is an important limitation for turbulence modelling of such flows. Moreover, the Smagorinsky model could over-predict SGS dissipation and modify the true energy cascade.
- The Smagorinsky coefficient should be adjusted for each turbomachinery flow field and this is carried out on the basis of proportionality between turbulent eddy viscosity and resolved velocity gradients. If the mean velocity gradients are considerable as compared with the turbulent ones, the turbulent viscosity is overestimated. The coefficient should be zero near

walls or in laminar flows. Implementation of a constant Smagorinsky coefficient in these regions overestimates SGS stress and may prevent flow transition to turbulence.
- Evaluation of the Smagorinsky model in a squirrel-cage fan shows weak correlations between modelled and real SGS quantities. These prediction inaccuracies may be due to misalignment of the SGS stress tensor and resolved strain rate tensor.
- The Smagorinsky coefficient is quite dependent on flow regime and filter size, especially in straining and de-straining conditions that are frequent in any turbomachinery flow. Estimation of the Smagorinsky coefficient on the basis of SGS dissipation indicates a diverse range of values for different flow conditions. The calculated Smagorinsky coefficient for squirrel-cage fans is considerably less than its classical value (0.16).

The Smagorinsky model is still used in various simulations, despite its shortcomings. Simplicity of its implementation and its well-behaved numerical characteristics are the main reasons for this popularity. Adjustment of the Smagorinsky coefficient according to local behaviour of the flow field may result in more accurate prediction of SGS stress. Two different approaches can be used to estimate the Smagorinsky coefficient:

1. Parameterization of the Smagorinsky coefficient as a function of local resolved-scale quantities.
2. Estimation of the optimal Smagorinsky coefficient during the simulation.

The dynamic Smagorinsky model (the latter approach) uses the smallest resolved scales to estimate the Smagorinsky coefficient both temporally and spatially. It assumes the scale-invariance hypothesis where the model coefficient is not dependent on the filter size and the calculated coefficient from resolved scales can be considered for SGSs. In comparison with the classical Smagorinsky model, the dynamic model is generally more accurate in predicting the eddy viscosity and turbulent energy flux. Despite popularity of the dynamic Smagorinsky model in flow simulations, it has shortcomings. The dynamic approach results in highly variable values of eddy viscosity (containing negative ones) that causes numerical instabilities. Implementation of an average in homogeneous directions of flow may avoid a fluctuating eddy-viscosity field. However, determination of homogeneity direction is not necessarily possible for the complicated three-dimensional flow in any type of turbomachine. Furthermore, scale invariance, which is the basic assumption of a dynamic model, would not be correct when the filter scale is far beyond the inertial sub-range. A scale-dependent dynamic model that determines the ratio of coefficients at two different scales may overcome this drawback. Performance of the scale-dependent dynamic model for eddy-viscosity prediction is superior to the standard dynamic model, although it includes a larger computational cost due to one more filtering operation.

The similarity model is another common SGS model that is postulated on the basis of similarity between the flow structures at scales just below and above the characteristic scale, Δ. The SGS stress tensor can be predicted from the resolved velocity field:

$$\tau_{ij}^{Sim} = C_{Sim}\, L_{ij} = C_{Sim}\left(\overline{\overline{u_i}}\,\overline{\overline{u_j}} - \overline{\overline{u_i}\,\overline{u_j}}\right) \tag{1.21}$$

where the lower and upper over-bar symbols denote the first and second filtering operations, respectively. The second filter is generally larger (usually, twice larger) than the first filter scale. Evaluation of the similarity model in various flow fields (including the squirrel-cage fan) shows that this model is more accurate than the Smagorinsky model. Unlike eddy-viscosity models, the similarity model is capable of predicting backscatter of energy.

The similarity model coefficient (C_{Sim}) is usually considered to be equal to or in the order of unity. However, an appropriate value for this coefficient depends on the implemented filter scale and flow characteristics. The similarity coefficient can be dynamically obtained during the simulation.

Although the similarity model is successful in predicting SGS stress, it is not sufficiently dissipative and underestimates SGS dissipation. This problem can be overcome by the mixed model that adds a dissipative eddy-viscosity term into the standard similarity model. Assessment of various models in prediction of SGS quantities shows that the mixed model is superior to the above-mentioned SGS models.

1.5.2 Computational considerations

1.5.2.1 Fan geometry in CFD simulations

Accurate boundary conditions and geometry should be considered when a fan is modelled in order to obtain valid results. Presence of separation and reverse flow in the centrifugal fans results in three-dimensional anisotropic flow downstream the blades. The numerical simulations based on two-dimensional geometries of the centrifugal fans show a higher pressure and efficiency as compared to experimental and three-dimensional numerical results. Two-dimensional models do not consider the inactive portion of the rotor and the separation zone behind the inlet, and therefore result in higher efficiencies. Figure 1.7 shows a common geometry in the numerical simulation of squirrel-cage fans. Different parts of the fan configuration in the numerical simulation include inlet semi-sphere, rotor, volute, straightener plates and outlet channel zones. The radius of the semi-sphere is at least four times larger than that of the inlet.

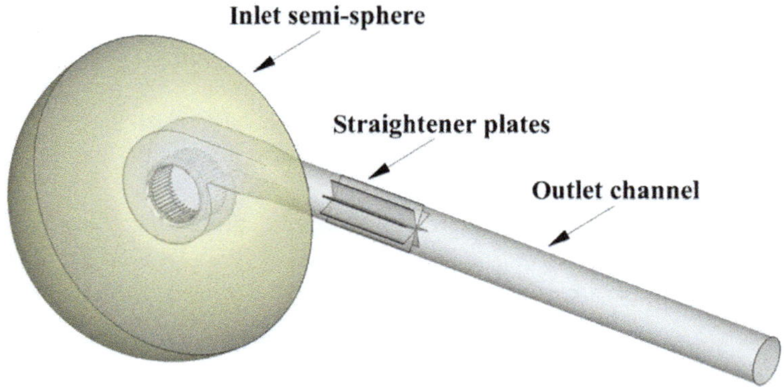

Figure 1.7 General configuration of the fan for numerical simulation.

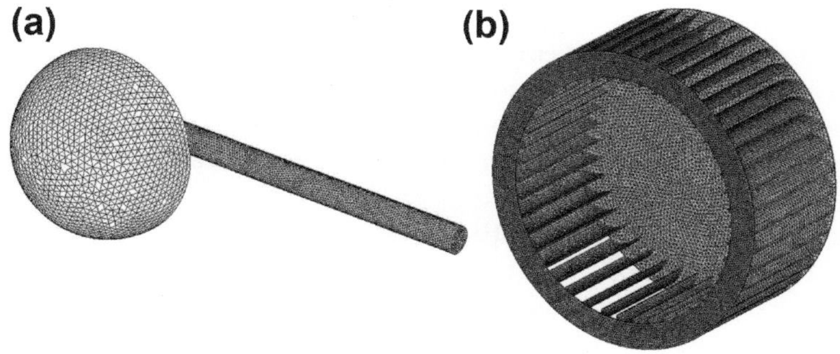

Figure 1.8 Sketch of the mesh: (a) on the walls of the considered geometry in numerical simulation; (b) on the rotor blades.

This is sufficiently large to remove any distortion. The role of a straightener in the outlet duct is to remove separation-induced vortical structures that can affect the measured pressure at the outlet duct. The outlet duct and the straightener are according to the ISO5108.

1.5.2.2 General considerations for the grid type

Numerical prediction of fan characteristics shows that the hexahedral and tetrahedral mesh topologies result in similar levels of accuracy. However, the tetrahedral mesh exhibits a more diffusive character than the hexahedral mesh. Therefore, for a detailed investigation of the complex flow field in the fans, the hexahedral mesh is a better choice.

The mesh should be refined near the volute cut-off, in the impeller domain, and around the straightener plates in order to resolve sufficient turbulent kinetic energy. A sketch of the suitable mesh on the outer walls of the geometry of Figure 1.7 is presented in Figure 1.8.

1.5.2.3 Boundary conditions

A common experience for determining fan performance is to simulate its steady-state flow field using the rotating reference frame as the impeller zone. The simulation should be performed at different flow rates. Therefore, at the initial execution for the maximum flow rate, the pressure outlet and pressure inlet boundary conditions are suggested for the surface of the semi-sphere (at the inlet) and the exit of the outlet duct, respectively. For lower flow rates, the mass flow inlet and outflow conditions are recommended as the boundary conditions on the inlet and outlet of the fan, respectively.

Unsteady simulations particularly in the case of LES require more attention to the inlet boundary condition. The characteristics of flow structures at this section are deterministic downstream and should be consistent with the dynamics of the inlet flow. Earlier studies with LES have often constructed an inlet boundary condition by adding random fluctuations to the mean inlet profile to generate sufficient turbulent kinetic energy for the inflow. However, absence of temporal or spatial correlation in these fluctuations results in noncoherent flow structures that are not sustainable after advection into the flow domain. In other words, the synthetically generated fluctuations have more energy in the high-wavenumber region, have a potential for rapid dissipation, and do not initiate the real turbulent structures. Despite these drawbacks, this method is still used in LES of various flows.

A reliable way to impose the realistic fluctuations into the inlet boundary condition is to extract the inflow condition from a spatial measurement technique such as PIV. The low temporal frequency of data acquisition can be compensated by implementing proper orthogonal decomposition (POD) or linear stochastic estimation (LSE).

Use of hybrid methods is also appropriate for generation of suitable boundary conditions at the inlet. The inflow condition can be obtained by extending the computational flow domain and combining RANS–LES approaches. Continuity of flow variables and their fluxes through the interface is the most important challenge in the hybrid approach.

1.6 Velocity measurement techniques and their considerations

A main prerequisite for any future analysis of a turbomachine is to extract its fluid velocity field. This includes study of turbulent-flow characteristics, identification of complex flow phenomena and optimization of geometry to enhance performance and reduce flow-induced noise.

The velocity measurement techniques could be classified into intrusive and nonintrusive techniques. The latter is superior because it causes no interference in the flow field and remotely measures the velocity field using buoyant particle tracers.

1.6.1 Intrusive measurement techniques

Pitot tube, hot-wire and hot-film anemometers are common measurement methods that extract point-wise velocity by direct contact between the anemometer probe and the flow stream. These methods affect the near-probe flow and are considered as intrusive measurement techniques.

The pitot tube extracts velocity after measuring total (stagnation) and static pressures. The difference between total and static pressures is the dynamic pressure ($\rho V^2/2$) which is a function of velocity at the measurement point. Although a pitot tube interferes with the near-probe flow field, its construction, setup and utilization are easy, and consequently it is widely used in performance tests of turbomachines to extract velocity, flow rate and pressure components.

Hot-wire and hot-film anemometers work on the basis of electrical resistance of their sensor (probe) that depends on heat transfer and therefore on the local flow velocity. The sensor is made of a thin metal wire (materials such as tungsten and platinum) or film so that its resistance is very sensitive to temperature variation. The probe is heated to a temperature that is considerably higher than the fluid temperature. The final temperature and resistance of the probe depends on the rate of heat loss from the probe into the fluid that strictly depends on the fluid velocity. Accordingly, measurement of voltage drop across the probe wire/film leads to extraction of local velocity magnitude.

In comparison with the pitot tube, the hot-wire and hot-film methods are more accurate and have considerably shorter response time (significantly higher measurement frequency). Consequently, when the purpose is to analyse temporal evolution of highly unstable and intermittent flow fields (such as turbomachinery flows), hot-wire and hot-film anemometers are superior to the pitot tube.

1.6.2 Nonintrusive measurement techniques

LDA and PIV are two common techniques that measure velocity components without interfering with the flow field. These methods use particle tracers (seeds) as the internal probes and measure velocity on the basis of light scattering of these fine tracers.

PIV and LDA complement each other. In comparison with LDA which is a pointwise velocimetry method, PIV provides the velocity field over a plane with appropriate resolution. LDA measures velocity with high frequency, while PIV, which is an image-based method, is a low-frequency measurement technique. Consequently, PIV is suitable for spatial analysis of flow structures, and LDA is appropriate for investigating temporal evolution of a flow field. Both of these methods are common in analysis of a complicated turbomachinery flow field that consists of a diverse range of temporal and spatial flow scales.

1.6.2.1 Laser Doppler anemometry

LDA, as is used in this book, is based on the Doppler shift of scattered light from a moving particle. The measurement point is a small volume of intersecting laser beams. In order to measure each velocity component, two laser beams are focused on the measurement volume. Intersection of laser wave fronts in the measurement volume (as shown in Figure 1.9) generates laser planes known as fringes. When a particle passes through the fringes, it scatters the laser light in all directions with different intensities (Mai scattering). A receiver collects the scattered light and focuses it on a photodetector. The frequency of received light is the Doppler shift between the incident and reflected light. This is proportional to the velocity component that lies in the plane of two incident laser beams and is normal to the fringes.

The spacing between laser fringes (δ_f) is calculated from

$$\delta_f = \frac{\lambda}{2\sin(\theta/2)} \tag{1.22}$$

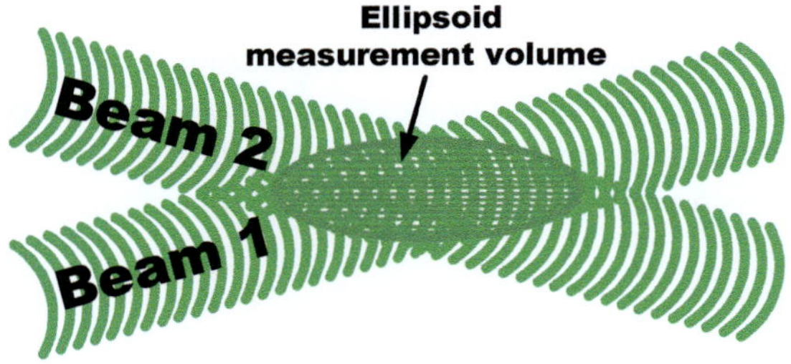

Figure 1.9 The measurement volume on the intersection of two laser beams; the fringes are shown by horizontal lines.

where θ is the angle between incident laser beams and λ is their wavenumber. After measuring the Doppler shift frequency (f_D) by the photodetector, the velocity component normal to the fringe plane is calculated from

$$u = f_D \delta_f = \frac{f_D \lambda}{2 \sin(\theta/2)} \tag{1.23}$$

Generation of three pairs of laser beams with different wavelengths in a single measurement volume results in three orthogonal components of velocity.

1.6.2.2 Particle image velocimetry

PIV measures the velocity field in a planar region on the basis of tracking seed location in two successive images of the same area with an infinitesimal time difference. Injected seeds are illuminated by a laser sheet and their location is captured by a high-definition double-frame charge-coupled device (CCD) camera (Figure 1.10). After detection of the displacement vector of particle tracers, the in-plane velocity vector is calculated by dividing the extracted displacement vector by the time interval.

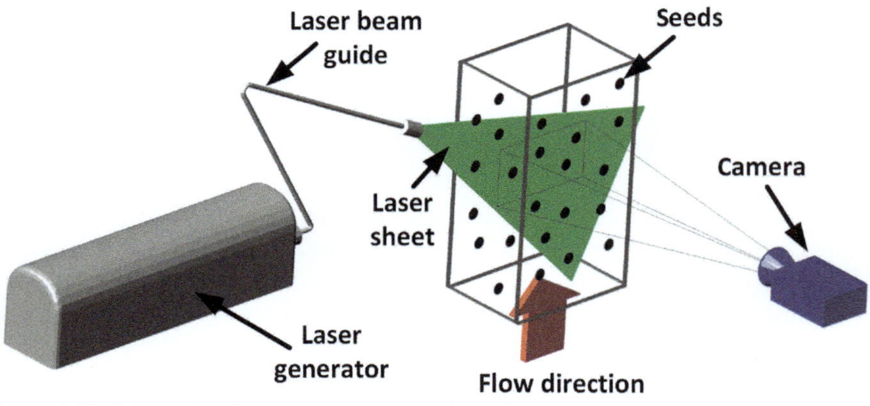

Figure 1.10 Schematic of a PIV setup that consists of a laser light generator, a laser guide/convertor system, a camera and seeding particles.

The main process in PIV is to determine the displacement vectors from raw black and white image pairs. There are numerous seeds in the measurement area, and the PIV processor is responsible for correct detection of particle displacement. A statistical image processing scheme should be applied to each pair of acquired images to determine the velocity vector map in the measurement area. This includes the following steps:

1. *Image subsampling*: The measurement area is first divided into a number of interrogation areas with selected dimension, namely 32 × 32 pixel, 64 × 64 pixel, and so on.
2. *FFT process*: The two-dimensional grey-scale maps are transformed into the Fourier space by applying the FFT. This transformation significantly reduces algorithm execution time.
3. *Cross-correlation procedure*: The cross-correlation of each pair of transformed light intensity functions is calculated after complex conjugate multiplication of these two functions.
4. *Inverse Fourier transformation and peak detection*: The obtained cross-correlation function in step #3 should be inversed in order to determine the maximum correlation in the image space and detect its corresponding displacement vector. A sub-pixel interpolation scheme is then applied to enhance the accuracy of the detected peak.
5. *Conversion of displacement to velocity vector*: The detected displacement vector in step #4 is simply converted into the velocity vector after dividing by the time interval between image frames.

The extracted velocity map from PIV measurement could be used for direct calculation of various flow quantities such as vorticity, strain rate tensor, turbulence stress tensors, various terms of the energy budget equation and so on. These quantities provide a basis for analysis of turbulent-flow structures, evaluation of turbulence models and perhaps enhancement of their performance.

1.6.2.3 Other types of PIV systems

There are other variations of the classic PIV system that are useful for complex applications. A summary of these PIV-based systems is as follows:

- *Stereoscopic particle image velocimetry (SPIV)*: This method uses two cameras that capture images through different viewing angles. Combination of the acquired image pairs provides the out-of-plane component of velocity in addition to the in-plane velocity components.
- *Dual-plane stereoscopic PIV*: The SPIV method only permits calculation of in-plane velocity gradients, because it is a single-plane velocimetry technique. The dual-plane stereoscopic PIV system utilizes two close laser sheets and four double-frame cameras to measure three velocity components at two adjacent measurement planes. The extracted data are consequently appropriate for calculation of complete velocity gradient components.
- *Tomographic PIV*: This method is capable of extracting a three-component velocity vector map in a three-dimensional measurement volume. The tomographic PIV technique utilizes several cameras to capture simultaneous images from different views of the illuminated measurement volume. A three-dimensional intensity field is reconstructed from these recorded images. The three-dimensional velocity vector map could be extracted by implementing a three-dimensional cross-correlation procedure on a pair of intensity fields (with a tiny time difference).

- *Micro-PIV*: This method uses an epifluorescent microscope to measure the flow in microscales. While the classic PIV method works on the basis of two-dimensional characteristics of the measurement plane, micro-PIV uses the capability of the objective lens to focus only on one plane and consequently creates a two-dimensional plane of observable particles.

1.6.3 Challenges of flow measurement in turbomachines

Complex and asymmetric geometry, rapid movement of rotor blades, nonuniform and anisotropic flow structures, and a diverse range of spatial and temporal flow scales are features that make turbomachinery flow measurement a challenging task. This section discusses the challenges and considerations in implementing LDA and PIV techniques to the three-dimensional flow of turbomachines.

1.6.3.1 Seeding the measurement area

The fluid velocity is calculated from seed displacement. Accurate following of fluid movements and rapid response to fluid fluctuations are essential features of correct seeding. Particles should be large enough to scatter sufficient light and yet small enough to accurately follow fluid motion.

Uniform and sufficient density of seeding particles in the measurement area is essential for accurate nonintrusive spatial measurements such as the PIV method and its derivatives. However, turbomachinery flow is significantly nonuniform where fulfilment of this condition is very difficult. It is a serious challenge for open turbomachinery setups but could be substantially resolved by utilizing a closed-cycle turbomachine and injecting sufficient seeding particles in the flow.

LDA measurements also suffer when seeding density is nonuniform. Repeated passage of tracer particles through the measurement volume with regular time intervals creates a high-quality signal for the LDA receiver that results in reliable data acquisition. Presence of multiple seeds in the measurement volume reduces the validity of extracted velocity components. Therefore, added seed intensity does not necessarily resolve the nonuniform seeding problem of turbomachinery flows.

1.6.3.2 Synchronization and encoding

The PIV processor synchronizes laser pulses with camera shots. Synchronization is required for turbomachinery flow when analysis of a particular orientation of the rotor blade is attempted. An encoder should trigger the PIV system and synchronize laser pulses and camera recordings with the target angular orientation of the rotor.

Rotary and optical encoders generate the required signal for turbomachinery measurements. Optical encoders are installed to observe a target sign on the rotor shaft or any location on the rotor surface, while the rotary encoders are directly installed on the rotor shaft. Indirect encoding of rotor orientation in turbomachines that are turned with the belt-pulley system is not reliable, because the slip between belt and pulley alters the relative orientation of rotor and electromotor shafts.

There are two types of rotary encoders, namely, absolute and incremental encoders. The absolute encoder maintains the previous angular position when the system is shut down. This encoder is appropriate for measurements at a particular orientation of the rotor (phase-resolved measurements). In contrast, the incremental (relative) encoder records relative changes in angular position, but does not preserve the previous rotor position.

The importance of accurate encoding and synchronization is a direct function of the number of rotor blades. Increasing the count of rotor blades reduces the blade-to-blade distance and consequently enhances the sensitivity of flow structures to rotor orientation.

1.6.3.3 Optical access, refraction and reflections

Nonintrusive measurement techniques require optical access to the measurement volume. This access is generally provided by constructing the casing of the turbomachine from transparent materials such as plexiglass. When the measurement area is located within the rotor/stator blades, these parts should also be transparent.

When the camera view is not normal to the measurement plane, refraction of laser light affects the acquired images in PIV-based methods. Camera calibration removes refraction-based distortions in images.

Reflection of laser light from the rotor/stator blades generates optical noise, if the incidence position is within the measurement plane. It also deteriorates the quality of received signal in LDA due to generation of additive scattered light after incidence of laser light with turbomachine blades.

The best optical condition for nonintrusive measurements can be provided when the turbomachine walls (casing and rotor/stator blades) and the operating fluid have the same refraction index. This is however practical only for closed-cycle turbomachinery flows.

1.6.3.4 Measurement settings

There are also other essential considerations and settings for a reliable nonintrusive measurement in the complicated flow of a turbomachine. Loss-of-pair is an important unfavourable effect in the PIV method that means some seeds are present in an interrogation area at the first image frame, but move to other interrogation areas at the second image frame. Laser light thickness, time interval between laser pulses and interrogation area dimensions are the PIV characteristics that should be appropriately adjusted to minimize this unfavourable effect and enhance data acquisition quality. A compromise between elevation and reduction of these three parameters determines the optimum setting for PIV measurement.

The turbomachinery flow has a wide variety of velocities. An appropriate time interval between laser pulses (or camera recordings) should be set to minimize loss-of-pair and pixel-locking effects in various regions of a measurement volume. Although reduction of this time interval decreases the possibility of loss-of-pair, it could not be very short, because its over-reduction increases pixel locking, particularly at regions with slower flows. Generally, high-speed flows require a short time interval while a larger time interval should be selected for low-speed flow regimes.

Dimension of an interrogation area should be determined after a compromise between higher spatial resolution and more reliable measurement. Implementation of a smaller interrogation area increases the spatial resolution of the extracted velocity field and consequently results in closer scales to the Kolmogorov length scale (the smallest dissipative scale of flow). On the other hand, several seeding particles should be located in each interrogation area for an acceptable cross-correlation procedure (typically 5−10 particles). Moreover, reduction of interrogation area size increases the possibility of loss-of-pair. The high-gradient flow fields such as the turbomachinery flows require a sufficiently large interrogation area that is capable of resolving a larger variety of seeding displacements.

Laser thickness depends on three-dimensional extent of flow at the measurement area. It is determined after a compromise between maximizing laser light intensity and minimizing loss-of-pair. Laser light intensity and therefore strength of illuminating the particle tracers is increased by utilizing a narrower laser sheet. However, the laser thickness should be large enough to minimize loss-of-pair, particularly for flows with considerable out-of-plane velocity components such as turbomachinery flow fields.

1.7 Final remarks

It is now clear that the squirrel-cage fan has applications in the ventilating industry but at the same time is an excellent instrument for fluid mechanics research. Numerical simulation of the fan is able to show general performance and fluid mechanic patterns and guide experimental research. On the other hand, experimental work is a reliable and potential tool for extracting global (measurement of head coefficient and efficiency), local (LDA and PIV) and directional (noise) characteristics.

Further reading

Akbari, G., Montazerin, N., & Akbarizadeh, M. (2011). Stereoscopic particle image velocimetry of the flow field in the rotor exit region of a forward-blade centrifugal turbomachine. *Proceedings of the IMechE, Part A: Journal of Power Energy, 226*, 163−181.

Ansys fluent 12.0, Theory guide. (April 2009).

Bardina, J., Ferziger, J. H., & Reynolds, W. C. (1980). Improved subgrid scale models for large-eddy simulation. In *Am. Inst. Aeronaut. Astronaut., fluid and plasma dynamics conference, Snowmass.*

Batten, P., Goldberg, U., & Chakravarthy, S. (2004). Interfacing statistical turbulence closures with large-eddy simulation. *AIAA Journal, 42*(3), 485−492.

Clark, R. A., Ferziger, J. H., & Reynolds, W. C. (1979). Evaluation of subgrid-scale models using an accurately simulated turbulent flow. *Journal of Fluid Mechanics, 91*(1), 1−16.

Chen, J., Katz, J., & Meneveau, C. (2005). Implication of mismatch between stress and strain-rate in turbulence subjected to rapid straining and destraining on dynamic LES models. *Journal of Fluids Engineering, 127*, 840−850.

Durst, F., Melling, A., & Whitelaw, J. H. (1981). *Principles and practice of laser-Doppler anemometry* (2nd ed.). Academic Press.

Germano, M., Piomelli, U., Moin, P., & Cabot, W. H. (1991). A dynamic subgrid-scale eddy viscosity model. *Physics of Fluids, 3*, 1760–1765.

Horiuti, K. (1997). A new dynamic two-parameter mixed model for large-eddy simulation. *Physics of Fluids, 9*, 3443–3464.

Kleissl, J., Parlange, M. B., & Meneveau, C. (2004). Field experimental study of dynamic Smagorinsky models in the atmospheric surface layer. *Journal of the Atmospheric Sciences, 61*, 2296–2307.

Li, N., Balaras, E., & Piomelli, U. (2000). Inflow conditions for large-eddy simulations of mixing layers. *Physics of Fluids, 12*, 935–938.

Liu, B., Yu, X., Liu, H., Jiang, H., Yuan, H., & Xu, Y. (2006). Application of SPIV in turbo-machinery. *Experiments in Fluids, 40*, 621–642.

Liu, S., Katz, J., & Meneveau, C. (1999). Evolution and modeling of subgrid scales during rapid straining of turbulence. *Journal of Fluid Mechanics, 387*, 281–320.

Liu, S., Meneveau, C., & Katz, J. (1994). On the properties of similarity subgrid-scale models as deduced from measurements in a turbulent jet. *Journal of Fluid Mechanics, 275*, 83–119.

Mason, P. J. (1994). Large-eddy simulation: a critical review. *Quarterly Journal of the Royal Meteorological Society, 120*, 1–26.

Meneveau, C., & Katz, J. (2000). Scale-invariance and turbulence models for large-eddy simulation. *Annual Review of Fluid Mechanics, 32*, 1–32.

Perret, L., Delville, J., Manceau, R., & Bonnet, J. P. (2006). Generation of turbulent inflow conditions for large eddy simulation from stereoscopic PIV measurements. *Journal of Heat Fluid Flow, 27*(4), 576–584.

Piomelli, U., & Zang, T. A. (1991). Large-eddy simulation of transitional channel flow. *Computer Physics Communications, 65*, 224–230.

Pope, S. B. (2000). *Turbulent flows* (1st ed.). Cambridge: Cambridge University Press.

Porté-Agel, F., Parlange, M. B., Meneveau, C., & Eichinger, W. E. (2001). A priori field study of the subgrid-scale heat fluxes and dissipation in the atmospheric surface layer. *Journal of Atmospheric Sciences, 58*, 2673–2698.

Raffel, M., Willert, C. E., Wereley, S. T., & Kompenhans, J. (2007). *Particle image velocimetry: A practical guide* (2nd ed.). Springer.

Schlüter, J. U., Pitsch, H., & Moin, P. (2003). Boundary conditions for LES in coupled simulations. In *41st aerospace sciences meeting and exhibit, AIAA, Reno, Nevada.*

Spalart, P. R. (1991). Strategies for turbulence modeling and simulations. *International Journal of Heat Fluid Flow, 21*, 252–263.

Speziale, C. (1991). Analytical methods for the development of Reynolds-stress closures in turbulence. *Annual Review of Fluid Mechanics, 23*(1), 107–157.

Smagorinsky, J. (1963). General circulation experiments with the primitive equations. I. The basic experiment. *Monthly Weather Review, 91*(3), 99–164.

Sullivan, P. P., Horst, T. W., Lenschow, D. H., Moeng, C. H., & Weil, J. C. (2003). Structure of subfilter-scale fluxes in the atmospheric surface layer with application to large-eddy simulation modeling. *Journal of Fluid Mechanics, 482*, 101–139.

The International Organization for Standardization. (2007). *International standard ISO5801: Industrial fans, fan performance testing using standardized airways.*

Vreman, B., Geurts, B., & Kuerten, H. (1994). On the formulation of the dynamic mixed subgrid-scale model. *Physics of Fluids, 6*, 4057–4059.

Inlet configuration

2

2.1 Why the inlet is important?

There are three distinct phenomena that are generally held responsible for poor efficiency of a forward-curved fan. The first is the flow separation that starts from the inlet and develops into the blades and then into the volute (Figure 1.3). It is stated that this occupies up to a third of the rotor width. Second, there is another separation mode that is common at the blade suction sides. This happens since the blade profiles of such fans are constructed from sheet metal with non-aerodynamic cross-sections. Larger inlet and exit angles are selected to inhibit flow reattachment on the blade suction sides. Still they usually experience a large angle of attack and a sharp turn. The third phenomenon is the reversing flow from the volute towards the inlet through the rotor. This is another consequence of inlet-induced separation. These three separation modes are due to different mechanisms that are interrelated. They intensify each other and give poor fan characteristics.

The inlet directs the flow into the rotor, and the volute collects the flow from the rotor. These two components, which are usually manufactured as one item, also serve to correct the flow field through the blades. There is a general agreement among researchers of forward-curved fans that uniform flow through the blades may only occur if, as the flow enters the rotor blades, it is simultaneously turned from axial to radial and undergoes a large deceleration. However, real cases show that the impeller is not correctly matched to the inlet flow and the volute casing does not promote a uniform blade loading.

Flow measurements in the rotor exit region show that due to a large separation zone there is little flow through the rotor near the fan inlet. In contrast, regions with sufficient through-flow are characterized by flow turning from axial (at the rotor inlet) to radial (at the rotor exit), and this is accomplished in the blade passage. This is a primary cause of the flow complexity that enhances loss and noise. If a fairly uniform radial flow can be induced by the fan at the blade entrance, there would be a decrease in energy loss within the rotor and therefore improvement in performance. The importance of inlet-induced flow is also emphasized in backward-curved centrifugal fans due to the significant effect of inlet configuration on the aerodynamic loss and the generated noise.

2.2 Bell-mouth inlet

It is common practice to use a bell-mouth for the fan inlet. Generally, there are three types of bell-mouth inlets: (1) the outward inlet, (2) the inward inlet and (3) the annular inlet. Figure 2.1 presents the geometry of these three types of bell-mouth inlets.

Developments in Turbomachinery Flow.

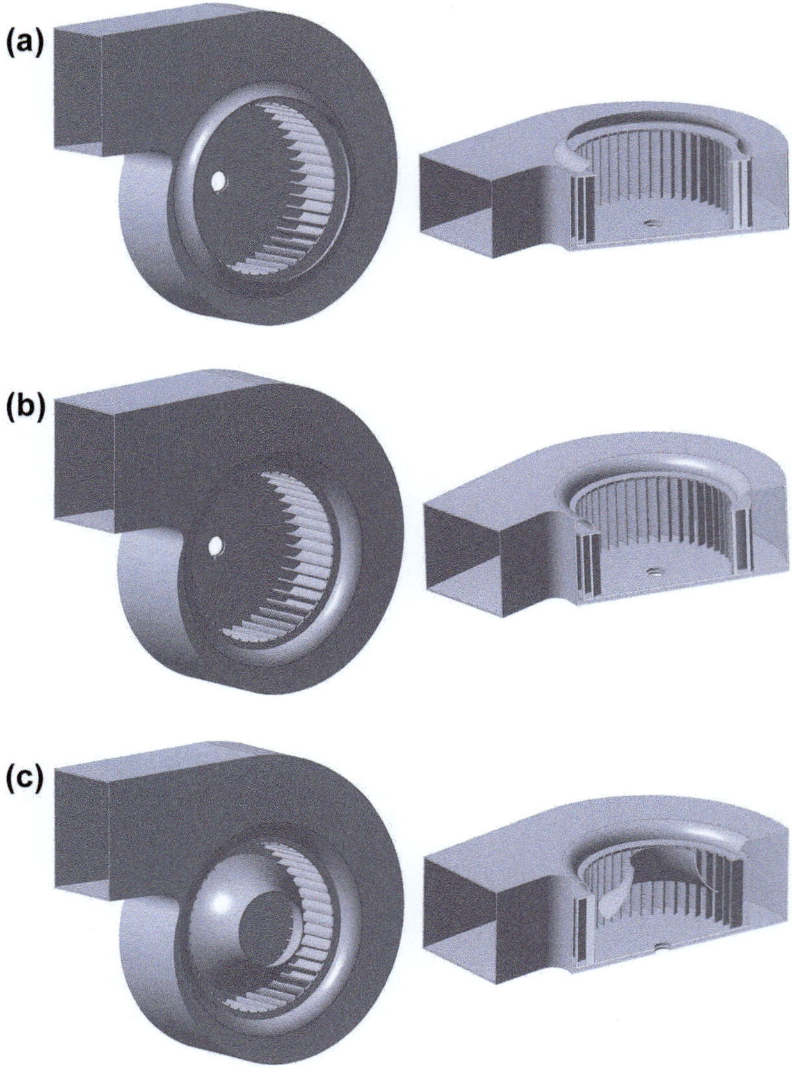

Figure 2.1 Different squirrel-cage fan inlets: (a) outward inlet; (b) inward inlet; (c) annular inlet.

The optimum inlet significantly increases the maximum flow rate through the fan and the maximum efficiency at the best efficiency point. The changes are so high that they outweigh all other attempts and geometric modifications to gain better performance from such fans. It is important to note that all optimizations and flow-field measurements on different fan geometries are only valid for that respective inlet configuration, which is often not even mentioned in the literature.

One solution to the inlet-induced separation problem is edge suction before the blades. This method attempts to suck out the vortices behind the inlet. An improvement

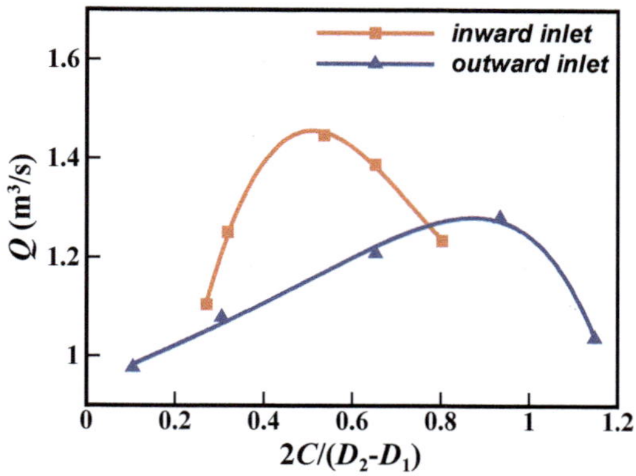

Figure 2.2 Maximum fan flow rate versus nondimensional gap spacing between the rotor and the inlet.

in efficiency is possible for specific combination of suction parameters. This was a theoretically successful but industrially impractical suggestion. The extra suction work makes the fan complicated and expensive.

Optimization of the gap spacing between the inlet and the rotor is another parameter that affects fan performance. Figure 2.2 examines the effect of gap spacing on flow rate of a sample fan for two different inlet configurations, namely, the inward and outward inlets. The horizontal axis is the nondimensional gap spacing that is defined based on the difference between rotor inner and outer diameters. Total pressure and efficiency show similar trends to curves of Figure 2.2 and are optimum for the same gap spacing. Smaller inlet gaps generate more uniform flow after the rotor while larger gaps create a recirculation zone behind the inlet. The flow rate is small for both a narrow and a wide gap spacing. It is not possible to generalize this trend for all fans, but an optimal setting must be measured for any individual fan design.

The flow may be guided into the rotor using a cone or an annular inlet that turns the flow towards the blade orientation (Figure 2.1(c)). The shape of the annular or the cone inlet affects characteristics of the recirculation zone behind the inlet. High curvature of streamlines at the inlet gap results in lower pressure at this region where more leakage ensues. Although it was expected that a larger guide diameter in the annular inlet would enhance performance, the opposite occurs in practice.

Examination of various inlet configurations shows that neither an annular or a cone inlet nor the reduction in the gap spacing enhances the fan performance significantly. The effect of outward and inward inlet configurations on fan characteristics is analysed in Section 2.4.

2.2.1 The shroud

The shroud is the ring that binds the near-inlet blade tips together (Figure 1.3). Although it is a part of the rotor, as far as performance is concerned, it acts hand in hand with the inlet. Therefore, it is appropriate to discuss it in this section.

There are optimum values for shroud dimensions. Narrower shrouds always result in larger mass flow rates and more uniform velocity profiles out of the rotor. The diameter of the optimum inlet is similar to the shroud inner diameter. This is an expectable issue because when the inlet diameter is smaller than the shroud inner diameter, blade suction area and active through-flow of the rotor are reduced. The opposite configuration (larger inlet diameter) causes the shroud surface to block the flow, and therefore it will further enhance separation and energy loss in the rotor. An even narrower shroud is preferable from a fluid mechanics point of view, and this is discussed in the following section.

2.2.2 Shroud-less rotor

Introduction of a new rotor which has no shroud is an important modification to the classical squirrel-cage rotor. This configuration only uses two rings that support the blades near their tips (Figure 2.3).

An inlet diameter, close to the shroud inner diameter, offers both the largest head and the largest volumetric flow rate. Using this inlet size (inner ring) for a shroud-less fan, the recirculation can freely occur from the volute into the blade tip. Such configuration activates the blade tip and enhances the entrainment of the reversing and the inlet flows. The Coanda effect helps a better turning of the incoming flow into the blades, and therefore a larger mass flow rate ensues.

Fan characteristics are not sensitive to the axial distance between the inlet and the shroud-less rotor when the rotor is close to the inlet. Increasing this distance causes a bistable phenomenon at the maximum flow rate. This is not usual in fans or even in other centrifugal turbomachines. Stall patterns usually occur at smaller flow rates. Bi-stable performance at larger flow rates is reported for squirrel-cage fans with no detail. Displacement of the rotor towards the inlet eliminates such instabilities.

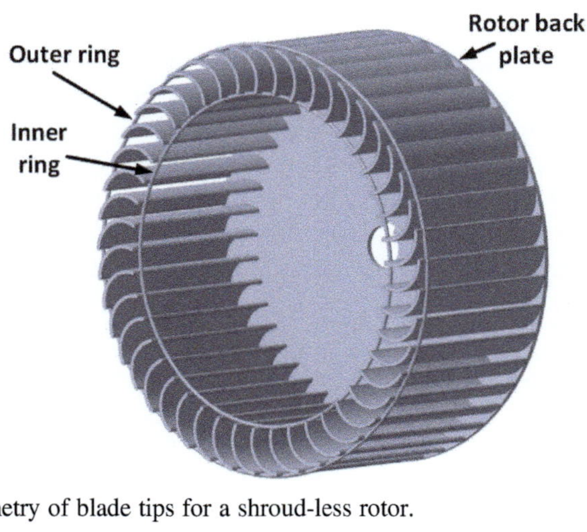

Figure 2.3 Geometry of blade tips for a shroud-less rotor.

2.3 Outward inlet

Once the flow has entered from an outward inlet (Figure 2.1(a)), rotor suction is to its advantage and directs the flow to the radial direction. It prevents separation, and therefore no vena contracta forms. Inlet curvature guides the flow towards the volute side walls, and therefore brings more of the rotor width into action. On the other hand, the inward inlet (Figure 2.1(b)) produces a highly separated and nonuniform flow after the inlet and across the width of the impeller.

Experimental evaluation of this configuration is performed by comparing the fan performance (in air) and the velocity measurement (in a water table) for four inlet lip geometries: one inward inlet, one flat inlet and two outward inlets (Figure 2.4). Rotational speed of the fan in the water medium is reduced an order of magnitude to repeat the minimum Reynolds number that was mentioned in Chapter 1. The water table is additionally deep enough to avoid the effect of surface waves on velocity profiles. The diameters of the inward and flat inlets are the average of the rotor inner and outer diameters. The outward inlets have two different diameters: one equal to the rotor inner diameter (small outward inlet) and the other equal to the rotor outer diameter (large outward inlet).

The outward inlet with a larger diameter is not as superior as expected but still shows a higher maximum flow rate and head. The improvement in head is marginal and starts to appear at flow rates over 50% of the maximum value. At lower flow rates, reversing flow and circulation round the rotor reduce the benefits of a streamline inlet.

The velocity profile of the same fan with the larger outward inlet lip is measured at three sections around the rotor, that is, $\theta = 180°, 270°, 360°$ (Figure 1.1). Whirl components are fairly uniform around the volute, and therefore are not shown here. This implies that irrespective of the behaviour of the radial components, the flow and vortices rotate uniformly round the rotor. Nondimensional radial velocity components for the first radial station on the three target sections are shown in Figure 2.5 (the velocity components are normalized by the rotor tip velocity at its outer radius, U_{tip}). Schematic inserts in each diagram show the angle where measurements are made

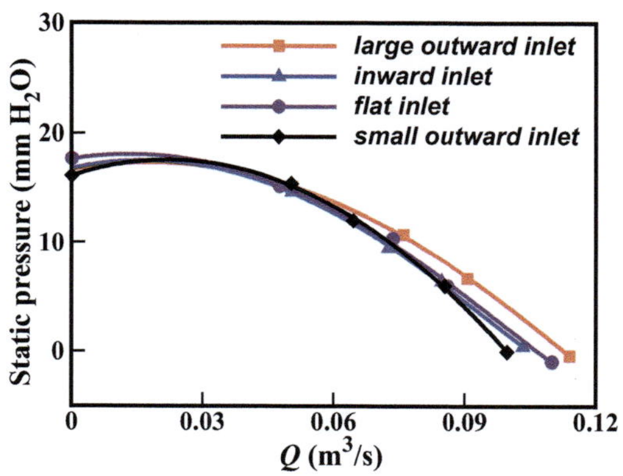

Figure 2.4 Characteristic curves for inward, outward and flat inlets.

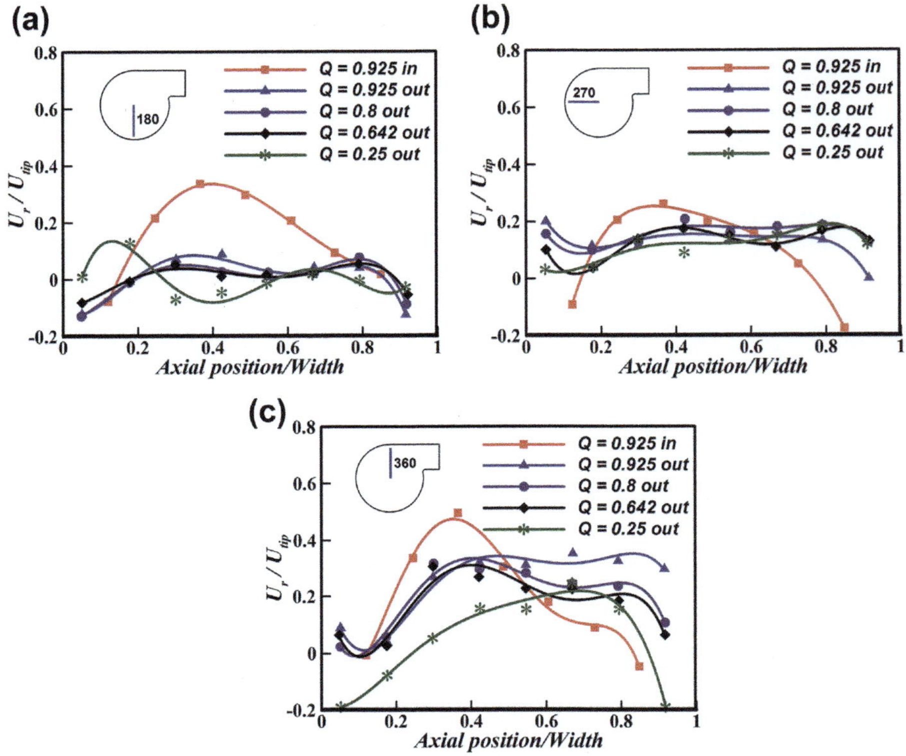

Figure 2.5 Nondimensional velocity for different normalized flow rates outside the rotor at nondimensional radius 1.09 at (a) 180°; (b) 270°; and (c) 360°. The terms 'in' and 'out' correspond to inward and outward inlets, respectively.

across the volute width. The flow rates in the legends are normalized with the maximum corresponding flow rate. A similar flow pattern with some decay is measured for other radial stations (not presented), and the possibility of local vortices inside the volute is therefore rejected. Measurements for the inward inlet lip on the same fan are also depicted for comparison.

Figure 2.5 shows that the nonuniform velocity profile for the inward inlet with a peak at around 40% of volute width is replaced with a uniform profile for the case of an outward inlet at all flow rates. Uniform blade loading and noise reduction are the main consequences of such even velocity profiles. The separated area that was reported behind the inward inlet (axial position/width → 1) is no longer present for the outward configuration. Figure 2.5 shows that the outward inlet has achieved its goal in controlling the flow turning and provides uniform blade loading.

Mass flux through the blades depends on angular positions. At 180° (Figure 2.5(a)) although the peak disappears for the outward inlet, the average velocity is almost zero. Apparently the reversing flow into the rotor that starts from the cut-off is still active. At 270° (Figure 2.5(b)) similar average velocities (but not similar profiles) for inward and outward inlets are achieved across the section. At 360° (Figure 2.5(c)) which has the

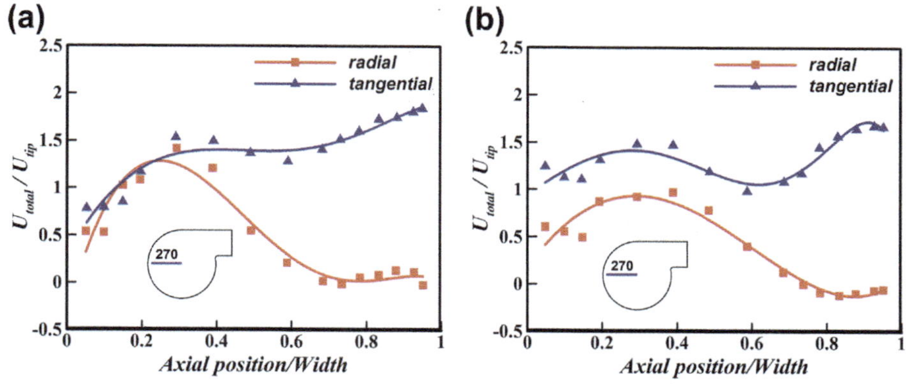

Figure 2.6 Radial and tangential components of the normalized velocity out of the rotor at $\theta = 270°$ for (a) inward inlet and (b) outward inlet.

largest volute cross-section, velocities are higher, but uniform flow is only present from 30% of volute width upwards.

Figure 2.6 shows both tangential and radial normalized velocity components for inward and outward inlets at the circumferential position of 270° of one forward-curved fan (as shown schematically in the figure). There is also a similar pattern at other locations away from the cut-off. Tangential components are larger than rotor peripheral speed. This is a theoretical characteristic of forward-curved fans. It is notable that there are locations across the width where the radial component of velocity is almost zero (i.e. no flow through the rotor) while high tangential velocities are observed. This flow has certainly gained its momentum from other axial locations in other circumferential positions. An even larger tangential velocity just behind the inlet is also worthy of note. There must be an axial transfer of flow and energy inside the rotor and the volute, towards the inlet. Highest tangential velocities just behind the inlet of squirrel-cage fans are a property that is often reported in different experiments.

The axial component of velocity is usually considered negligible as compared with tangential and radial components. Although this might be true on average, the local or instantaneous axial component of velocity is not negligible and causes flow deviations from two dimensional structures. The low radius of curvature of streamlines behind the inlet and the resulting axial pressure gradient inside the volute are simple justifications for such a phenomenon. This pattern also causes the nonuniform velocity profile at the fan exit.

Figure 2.6(b) shows that the two components of velocity behave similarly for outward inlets. This means that reductions in the radial component are followed by similar reductions of the tangential velocity. When the two components of velocity follow each other, the absolute velocity angle is uniform across the width.

Figure 2.7 shows the angle of absolute velocity (α) and the angle of relative velocity (β) in the r–θ plane for inward and outward inlets for maximum flow rate at three circumferential sections, namely, $\theta = 180°$, 270°, 360°. There is a general trend in both fans: where through-flow from the rotor into the volute exists, the absolute flow angle is more uniform for the outward inlet. The difference between absolute

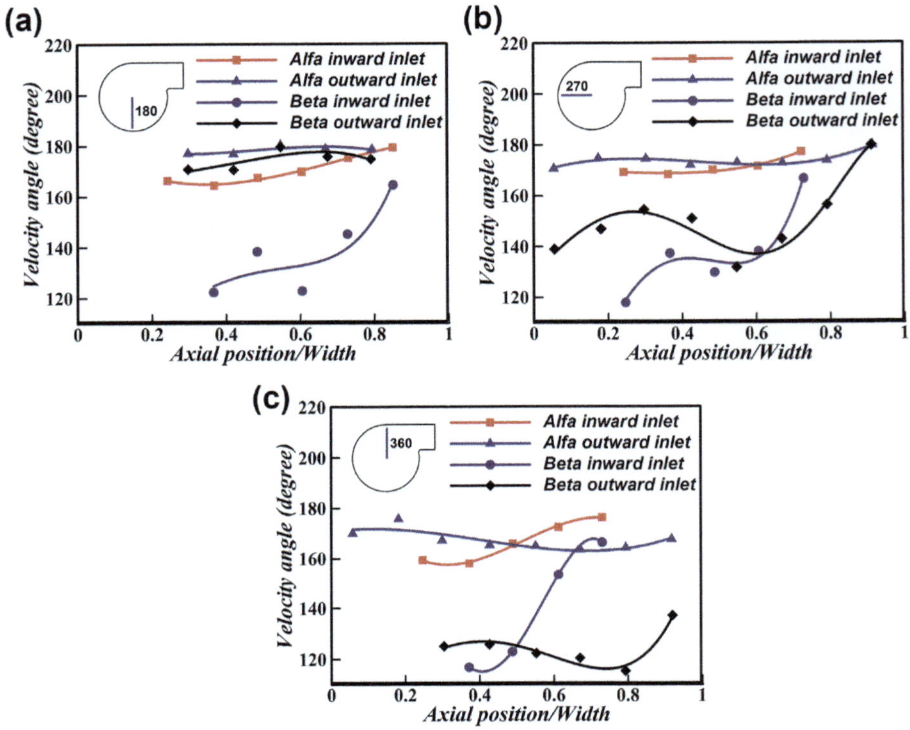

Figure 2.7 Absolute and relative velocity angles out of the rotor of fans with inward and outward inlets at (a) $\theta = 180°$; (b) $\theta = 270°$; and (c) $\theta = 360°$.

and relative angles is almost the same across the fan for the outward inlet. This means that blade action across the rotor is uniform for the outward inlet. This is not the case for the inward inlet.

A final point in Figure 2.7 is that the absolute velocity is tangent to the rotor near the inlet. This happens because the radial component of velocity diminishes there and therefore little, if any, through-flow exists. This topic is further explained in Chapter 3 where slip factor is studied.

The overall view from the angular flow distribution is that improving the inlet zone by changing the lip geometry necessitates a new optimum rotor and volute design. A smaller ratio of width to inlet diameter is needed to displace the same flow. The criterion $b/D_1 = 0.4$ allows for larger rotor inlet area (cylinder) than the fan inlet area (circle) to compensate for the separation vortex. This would possibly change in the optimization process for a lower value to bring the ratio of fan inlet area to rotor inlet area closer to unity. The optimization also reduces the redundant areas beside the volute back wall (axial position/width $\rightarrow$ 0).

The power spectral density function of both velocity components has a peak in the blade passing frequency at almost all measuring points for the outward inlet. It is stronger for the outward inlet and is detected over a wider width. This means that with an outward inlet, the entire rotor width experiences outflow. Other disturbances are smoothed out, and the passing blade is the dominant cause for unsteadiness.

The outward inlet is a possible alternative to the inward inlet. This inlet has some very specific characters that allow further insights into mechanisms of inlet-induced flow in forward-curved fans. A fan with the outward inlet has the advantage that it is less sensitive to the gap spacing between the inlet and the rotor. These results are not general to all fans, and therefore there are other geometrical parameters that could match the outward inlet for improved performance.

2.4 Final remarks

The studies on inlet optimization are all valid as long as the inlet geometry is a critical fan aspect. It is shown in Chapter 4 that volute angle controls the flow in the fan and overrides other geometrical parameters. Larger volute angles from 7° to 11° increase fan flow rate and reduce inlet effects.

Further reading

Denger, G. R., & McBride, M. W. (1990). Three-dimensional flow field characteristics measured in a forward-curved centrifugal blower using particle tracing velocimetry. In *Proceedings of fluid measurement and instrumentation forum, 95* (pp. 49–56). ASME Publication FED.

Montazerin, N., Damangir, A., & Kazemi Fard, A. A. (2001). Study of slip factor and velocity components at the rotor exit of forward-curved squirrel cage fans, using laser Doppler anemometry. *Proceedings of the Institution of Mechanical Engineers Part A: Journal of Power and Energy, 215*, 453–463.

Montazerin, N., Damangir, A., & Mirian, S. (1998). A new concept for squirrel-cage fan inlet. *Proceedings of the Institution of Mechanical Engineers Part A: Journal of Power and Energy, 212*, 343–349.

Montazerin, N., Damangir, A., & Mirzaei, H. (2000). Inlet induced flow in squirrel-cage fans. *Proceedings of the Institution of Mechanical Engineers Part A: Journal of Power and Energy, 214*, 243–253.

Raj, D., & Swim, W. B. (1981). Measurements of the mean flow velocity and velocity fluctuations at the exit of an FC centrifugal fan rotor. *Journal of Engineering for Gas Turbines and Power, 103*, 393–399.

Roth, H. W. (1981). Optimierung von trommelläufer-ventilatoren. *Strömungsmechanik und Strömungsmaschinen, 29*, 1–45.

Samarbakhsh, S. (2006). *Study of volute profile in squirrel-cage fan* (MSc. thesis). Amirkabir University of Technology.

Rotor

3

3.1 Half-cone rotors

The idea of a half-cone rotor emerged from the fact that it is impossible to turn the flow 90° after the inlet: so why not align the rotor with the inlet flow? Figure 3.1 shows a cylindrical rotor beside two positive ($\alpha > 0$) and negative ($\alpha < 0$) half-cone rotors. The blade tip radius of a positive half-cone rotor decreases from the rotor inlet towards its backplate (Figure 3.1(b)). An opposite trend holds for the negative half-cone rotors. The benefit of a negative half-cone rotor is that a larger portion of flow passes through the blades further away from the inlet. A larger diameter in that region would transfer more energy to the flow and produce larger work.

These half-cone rotors share a common average diameter and a common rotor width, hence the average distance between the rotor outer diameter and the cut-off would be the same. This brings no change to the volute, but the inlet should be modified to match the rotor inlet diameter.

3.1.1 *Numerical evaluation of half-cone rotors*

Three-dimensional simulation of flow field helps to investigate the detailed behaviour of the squirrel-cage fan with different half-cone rotors. In this way, the fan characteristics plus its velocity and pressure fields are examined.

Figure 3.2 shows numerical data obtained for performance curves of fans with half-cone and cylindrical rotors. The integers in the legend of this and the following figures denote the angle of half-cone rotors (the 0 angle indicates the cylindrical rotor). Figure 3.2(a) depicts the total pressure coefficient versus the flow coefficient. Rotors with negative cone angles produce higher heads. The rotor with angle $\alpha = -10°$

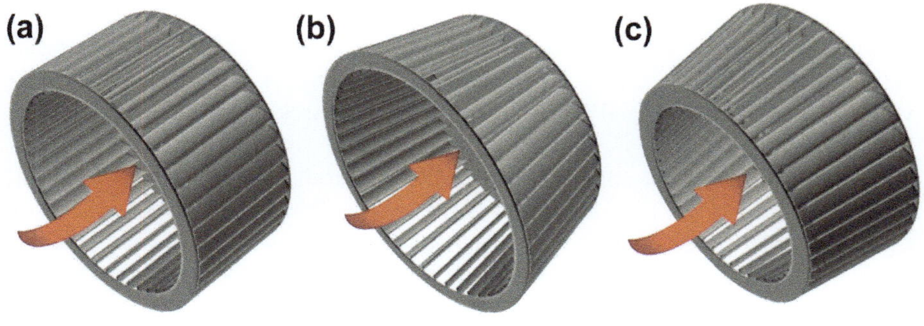

Figure 3.1 Geometry of different types of rotor: (a) cylindrical rotor ($\alpha = 0$); (b) positive half-cone rotor ($\alpha > 0$); and (c) negative half-cone rotor ($\alpha < 0$). The inlet flow is shown by arrows.

Developments in Turbomachinery Flow.

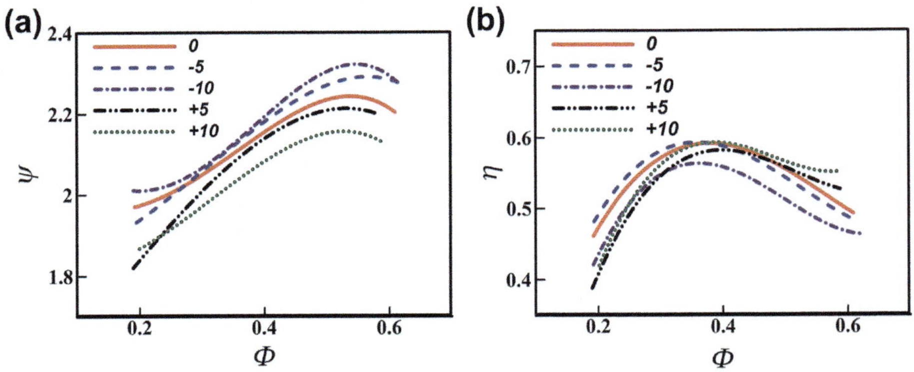

Figure 3.2 Comparison of fan performance characteristics for different rotor geometries: (a) pressure coefficient and (b) efficiency.

has the maximum pressure coefficient, while the rotor with angle $\alpha = +10^\circ$ generates the minimum pressure coefficient almost throughout the entire flow range.

Figure 3.2(b) compares the efficiency of different rotor geometries. The rotors with $+10^\circ$ and -10° cone angles have the highest and lowest efficiencies, respectively. It is a common pattern in turbomachines that higher load coefficients appear for lower efficiencies and vice versa. The famous Smith diagram for axial turbines is an example that shows the trade-off between these two items. This also happens in other turbomachines for various blade turning angles. Highly separated flow in forward-curved fans makes the blade turning angle a less-sensitive parameter. Half-cone rotors activate the same trade-off, and the selection would depend on the importance of the load or the efficiency for the application.

The flow field in the volute of fans with conical rotors has indicators that justify the trend of higher load coefficient for negative half-cones and higher efficiency for positive half-cones. Since the maximum pressure coefficient of all rotors is obtained at an approximate flow coefficient of 0.55, the analysis focuses on this flow rate.

The flow travels a complex path inside the volute. A typical flow pattern in the meridional plane is shown in Figure 3.3. A vortex forms just behind the inlet and causes a larger mass flux through the rotor towards the backplate. Static pressure across the rotor is an indicator for the reverse flow (that the vortex pushes from the volute towards the inlet) and the outward flow that passes through the blades away from the inlet.

Figure 3.4 presents a comparison of static pressure inside (at $r/r_0 = 0.65$) and outside (at $r/r_0 = 1.17$) of the rotors with different half-cone angles at $\theta = 0^\circ$ (r_0 is the outer radius of the rotor). It is expected in forward-curved fans that the flow passage through the rotor blades causes a reduction in static pressure which is coupled with an increase in velocity. The difference in the distribution of static pressure inside the five rotors in Figure 3.4 shows that rotors with positive angles have a uniform negative pressure gradient across the rotor. In contrast, negative half-cone rotors are characterized by reverse suction around the inlet, which accounts

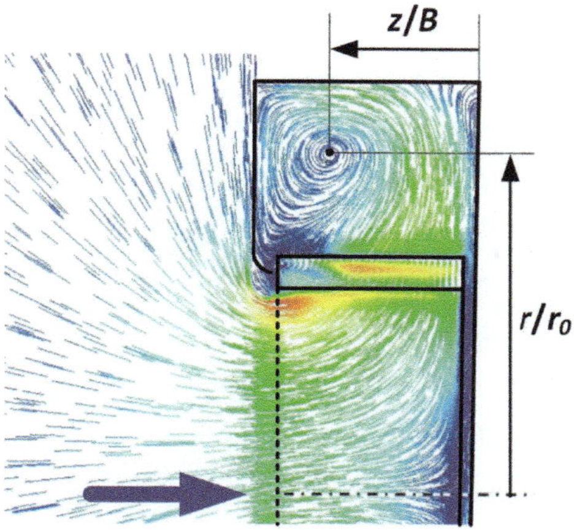

Figure 3.3 General pattern of the flow in the meridional plane of fan. The axial and radial distances are measured relative to the volute backplate and the rotor axis, respectively.

for the return flow from the volute to the inside of the rotor. Larger inactive area of the rotor and smaller efficiency are the main consequences for such reverse flow in negative half-cone rotors.

The circulation in the volute space (Figure 3.3) causes energy loss. The location of the centre of the vortex is shown in Figure 3.5 in terms of the distance from volute backplate (z/B) and the radial distance from the fan axis of rotation (r/r_0) which are both indicators for the vortex half size. The locations behind various conical rotors are obtained in three angular sections, namely, 270°, 180° and 360°. Figure 3.5 shows that the radial position of the vortex is the same for different cones in each angular section. However, axial distance is different. The change is more pronounced near the volute exit ($\theta = 360°$). All vortices have almost the same size at $\theta = 180°$, while the largest vortices are at $\theta = 270°$. The $+10°$ cone has a relatively smaller growth which makes it even smaller than the vortex for the cylindrical rotor. At $\theta = 360°$ and near the exit where there is the largest flow rate, cylindrical and positive rotors have vortices of the same size, but vortices behind negative cones are larger. The overall conclusion is that the vortex is smaller behind positive cones, which also justifies its higher efficiency.

The circumferential distribution of the radial velocity out of rotor can be a criterion for efficiency. When more flow leaves the rotor near the outlet, it travels a shorter distance in the volute with smaller loss. Generally, the maximum radial velocity in squirrel-cage fans is representative of the maximum flow rate that occurs near the volute outlet. Figure 3.6 shows the radial component of the normalized velocity in the fan with cylindrical rotor at different axial distances from the back plane of the volute at $r/r_0 = 1.17$. The maximum radial velocity occurs at $\theta = 25°$. There is reverse flow near the cut-off ($\theta = 50°$) with minimum mass flux through the blades. The radial velocity is negative in this region, which means that there is return flow from the volute into the blades. The distance from $z/B = 0.7$ to $z/B = 1$ is the inlet region, where radial

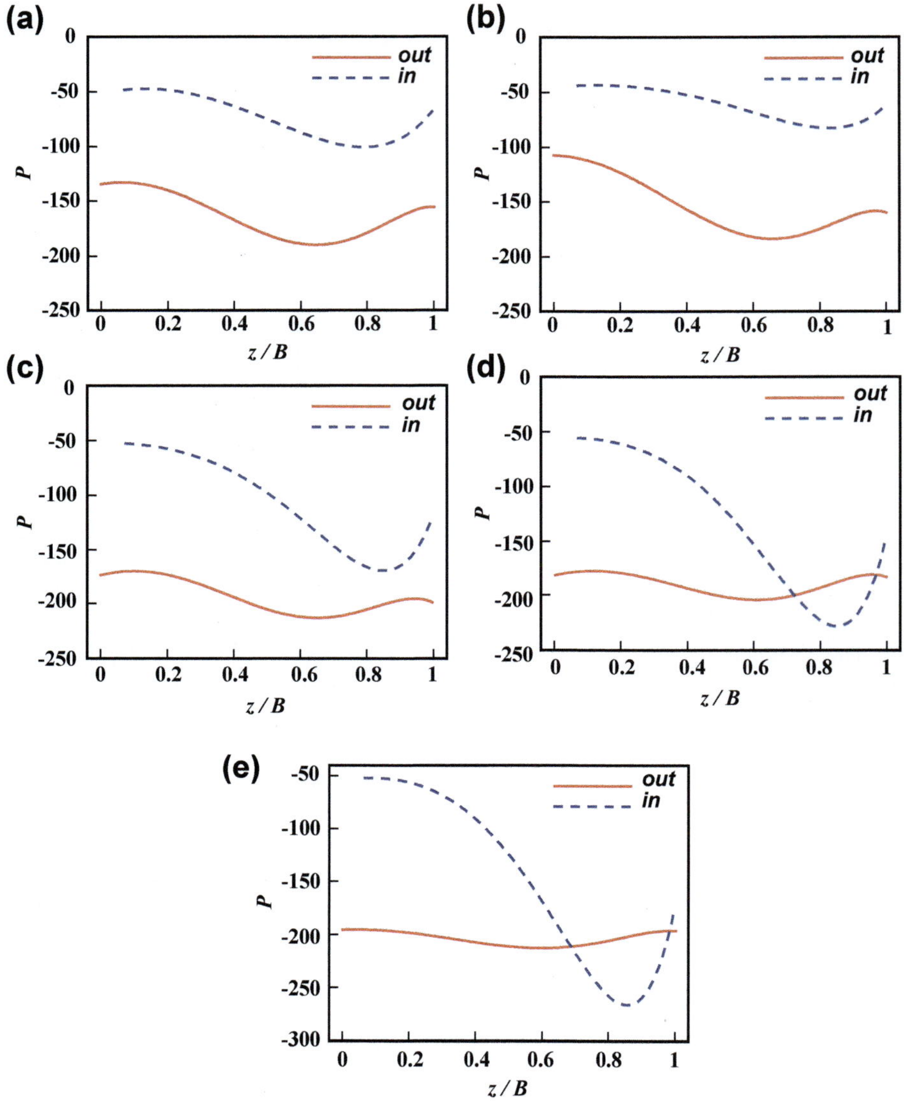

Figure 3.4 Static pressure inside (at $r/r_0 = 0.65$) and outside (at $r/r_0 = 1.17$) of the rotor at $\theta = 0°$. Measurements are for different conical angles: (a) $\alpha = +10°$; (b) $\alpha = +5°$; (c) $\alpha = 0°$ (cylindrical rotor); (d) $\alpha = -5°$; and (e) $\alpha = -10°$.

velocities are minimum. After this region, radial velocities increase in the active part of the rotor and reach a maximum at $0.2 < z/B < 0.3$.

Figure 3.7 shows the radial velocity for different rotors at angular positions of 25° and 270°. The area under radial velocity curve at 25° is divided by the respective area under the curve at 270° to present a general comparison between radial velocities at these two sections. This ratio is listed in Table 3.1 for different rotors. The +10° rotor

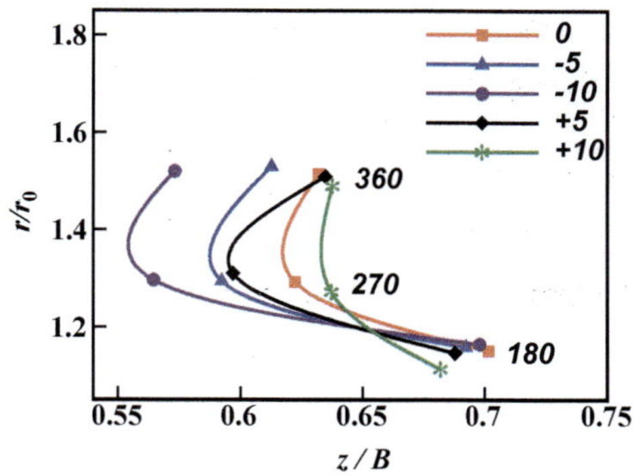

Figure 3.5 Location of the centre of vortex for different rotors at angular sections 180°, 270° and 360°.

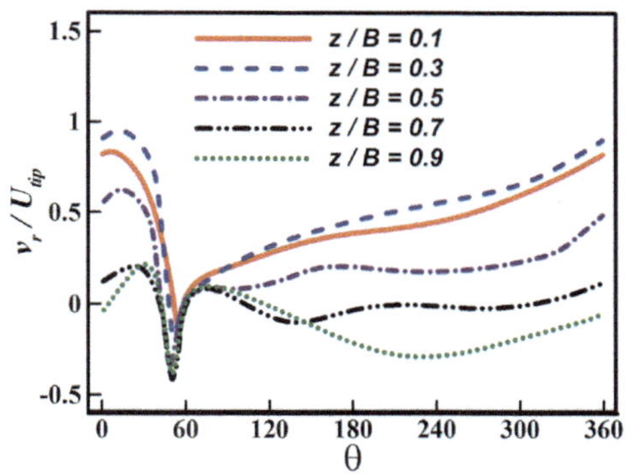

Figure 3.6 Radial normalized velocity out of the cylindrical rotor at $r/r_0 = 1.17$ for different axial locations.

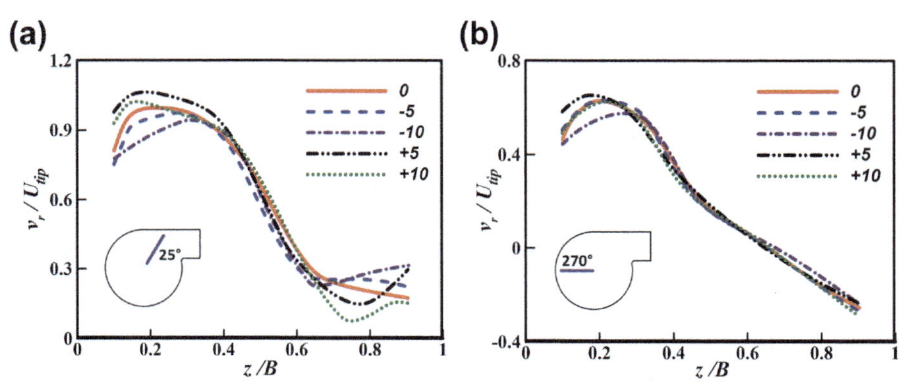

Figure 3.7 Axial variation of the radial velocity component for different rotor geometries at (a) $\theta = 25°$ and (b) $\theta = 270°$.

Table 3.1 The ratio of area under radial velocity curves at 25° and 270° for different rotor geometries

Rotor angle	$\frac{\text{Area under radial velocity curves at } 25^{\circ}}{\text{Area under radial velocity curves at } 270^{\circ}}$
0	2.8323
+5	2.8215
+10	2.9273
−5	2.7263
−10	2.7513

has the highest ratio, which means larger portions of the flow leave the rotor at 25° with smaller ensuing losses. Although the ratio is marginally larger for the positive half-cone rotors, differences are not high enough to be conclusive for their better efficiency.

Figure 3.8 shows the tangential velocity for different rotors at angular sections 25° and 270°. The rotors with negative cone angles have a higher tangential velocity away from the inlet. The reason for this behaviour is the larger diameter of these rotors at this region that transfers a proportionately large angular momentum to the flow. This pattern confirms the higher load coefficient of the negative rotors. Figure 3.8(b) demonstrates that the tangential velocity behind the inlet is considerably large for all rotors.

The flow through the cut-off clearance of a centrifugal turbomachine does not leave the volute and can reduce the machine efficiency. This part of the volute is the plane between the volute tongue and the rotor. Table 3.2 presents the mass flow across this surface for fans with different rotor cone angles (the rotor-to-tongue surface area of these fans is the same). The fan with +10° rotor has the smallest return flow back to the volute that supports the higher efficiency of this rotor.

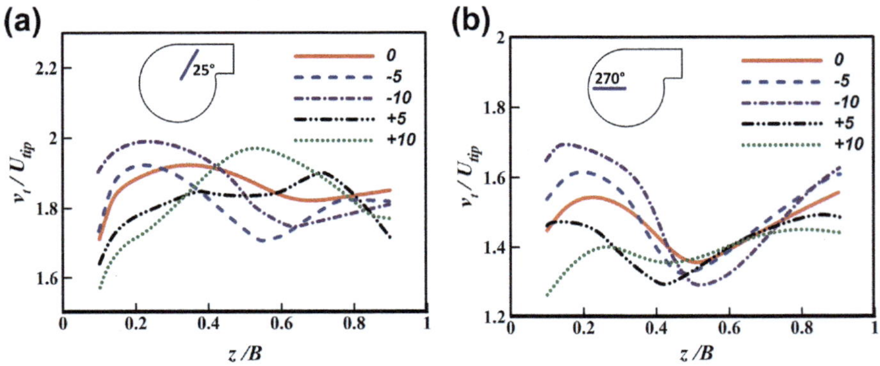

Figure 3.8 Axial variation of the tangential velocity component for different rotor geometries at (a) $\theta = 25°$ and (b) $\theta = 270°$.

Table 3.2 Calculated reverse mass flow rate crossing the tongue region for different rotors

Rotor angle	Mass flow rate (kg/s)
0	0.1431
+5	0.1405
+10	0.1388
−5	0.1444
−10	0.1449

A comparison of the results in Figures 3.2−3.8 as well as Tables 3.1 and 3.2 shows that smaller vortex behind the volute, less flow through cut-off clearance and larger active area of the rotor are all indicators for the higher efficiency of the positive half-cone rotors as compared to the others.

3.1.2 Experimental investigation of positive and negative cone angles

Experimental data on conical rotors for squirrel-cage fans in the range −10° to +10° are available for five different rotor angles with 5° increments (a few rotors have lean angles, that is, their blades are not aligned with the rotor axis). Figure 3.9 shows experimental performance characteristics of these rotors. All rotors in Figure 3.9(a) and (b) are constructed using similar blades with circular arcs. The selection of the range for cone angles is such that there would be no need to change the cut-off or the scroll casing. A common rotor width for all five cases means larger blade lengths for conical rotors. This is compromised since fan performance is viewed in terms of rotor and

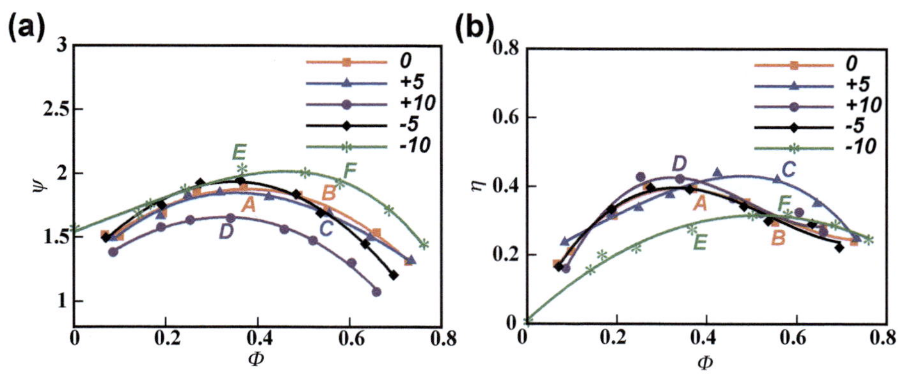

Figure 3.9 Experimental performance characteristics of half-cone rotors with different angles: (a) pressure coefficient versus flow coefficient and (b) efficiency versus flow coefficient.

scroll overall dimensions. Blades are of the same shape and their inlet and outlet angles are the same for all rotors and are set in their optimum configuration. Performance curves in Figure 3.2 do not show any of the optimum cone angles in Figure 3.9. This mismatch can only be explained later when the idea of lean angle is introduced.

Figure 3.9(a) shows that the pressure coefficient for the $-10°$ conical rotor is higher than that of the cylindrical rotor. This is not true for the cone angle of $-5°$ that also has a similar head and efficiency as the cylindrical rotor. According to fan efficiency curves in Figure 3.9(b), the efficiency of the $-10°$ rotor in the case of fairly small flow coefficients (point E) is less than the corresponding values for the cylindrical rotor (point A), while the opposite is true for flow coefficients that are more than 0.5 (point B for cylindrical and point F for conical rotors, respectively). Since fans normally operate in the right-hand part of the curves, production of a higher head for similar efficiency at larger flow rates would count as an advantage of the rotor with $-10°$ cone angle.

Pressure coefficient and maximum flow rate of $+10°$ conical rotor are smaller than the corresponding values for the cylindrical rotor. The efficiency of the $+10°$ rotor is not different in the operating flow coefficients. No difference is evident between the $+5°$ rotor and the cylindrical rotor as far as the pressure coefficient curve is concerned. However, the efficiency of the $+5°$ conical rotor at the flow coefficient 0.55, which is well inside the normal operating range of fan (point C), has improved up to about 40% relative to the cylindrical rotor (point B).

The overall conclusion is that inside the normal operating range of fan, the $-10°$ conical rotor has a larger head with no loss in efficiency and the $+5°$ rotor has an improved efficiency with no loss for head or flow rate. It is not evident from these experiments why there are optimum angles and why they are different between positive and negative cone rotors.

3.1.3 Flow field on the basis of velocity measurements

Measurement of velocity profiles reveals corresponding behaviours between velocity fields and performance curves. Primary velocity measurements for the five considered rotors in this section may show the trends and guide for possible future experiments with more details.

Three guidelines were followed in selecting the test points on the performance curves that included points A to F in Figure 3.9:

1. Flow fields for cylindrical, positive and negative conical rotors at their maximum respective head, namely, points A, D and E, respectively, reveal why the negative half-cone rotor has the largest head coefficient. This cone also experiences a notable efficiency drop at point E that recovers for larger flow rates.
2. Points B, C and F cover the relationship between efficiency and velocity field. A positive half-cone rotor shows a considerable improvement in efficiency with no loss of head. The efficiency for the negative half-cone rotor has also marginally improved as compared to the cylindrical rotor.
3. Points A and B and similarly Points E and F are on the same respective performance curves. A comparison between their flow fields could show the influence of flow rate and throttling on the velocity components.

A general knowledge is required about flow-field pattern at different levels of throttling in order to select angular sections for the velocity measurements. When the throttle is fully open, air flows out of the rotor at almost all circumferential locations. If the fan is throttled, a reversing flow into the rotor starts from around the cut-off. With further throttling, this reverse flow occupies a larger portion of the rotor perimeter. Velocity measurements at two circumferential positions, namely, $\theta = 270°$ and $\theta = 360°$, are available. These locations have outward flow for almost all points on performance curves. There are no data at other circumferential positions since in forward-curved squirrel-cage fans, small volute cross-sections at $\theta = 90°$ and $\theta = 180°$ allow a smaller portion of the flow out of the rotor, even at fully open throttle. Research on squirrel-cage fans has shown that the velocity profile out of the rotor is characterized by separation just after the inlet nozzle and the velocity profile through the rotor. Such profile that consists of jet and wake patterns, decays inside the volute and demonstrates major features of the fan flow field. The intention here is to present the velocity profiles as close to the rotors as possible. Hence, velocity measurements are at one radial position for the cylindrical rotor and two radial positions for half-cones.

3.1.3.1 Half-cone and cylindrical rotors at their maximum respective heads

Figure 3.10 shows the results of the velocity profile measurements for cylindrical ($\alpha = 0°$) and conical ($\alpha = +10°$ and $\alpha = -10°$) rotors at operating conditions A, D and E (Figure 3.9) and at $r/r_0 = 1.17$. Measurements are in sections $\theta = 270°$ (Figure 3.10(a)) and $\theta = 360°$ (Figure 3.10(b)). Inserts provide visual aid to distinguish between measurement sections. The z direction originates from the backplate. Since the separation zone after the inlet occupies about one-third of the rotor near $z/B = 1.0$, some sharp changes in the middle and the right side of curves (in Figure 3.10) are expected.

Both the circumferential and the radial components of velocity for a $-10°$ conical rotor at $z/B < 0.5$ (away from the inlet) are larger than the corresponding curves for the

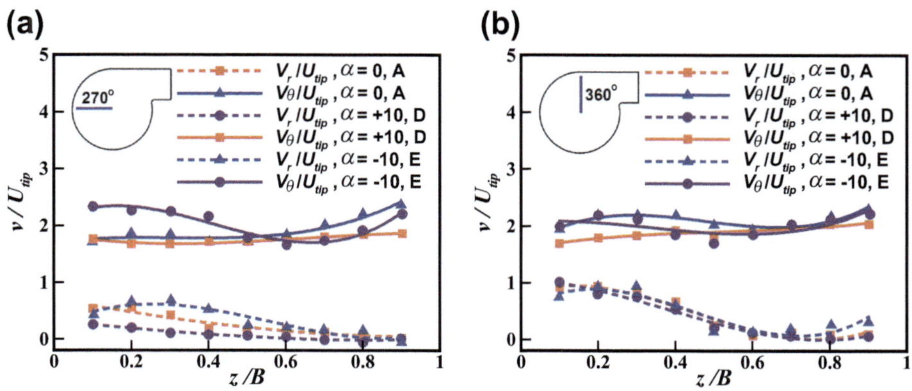

Figure 3.10 Axial variation of radial and circumferential velocity components in $r/r_0 = 1.17$ for three different rotors and two measurement sections: (a) $\theta = 270°$ and (b) $\theta = 360°$.

other two rotors at both sections. This is justified from the Euler equation for forward-curved blades. Since the radius is larger than average at $z/B < 0.5$, the main part of the flow passes through this rotor there, and this results in both a larger circumferential velocity and a corresponding radial component. The larger head for the entire head coefficient curve then ensues.

Figure 3.11 shows laser Doppler anemometry (LDA) measurements of the radial velocity at different angular positions of the exit region of a $+10°$ half-cone rotor at $r/r_0 = 1.17$ and $z/B = 0.3$. The maximum radial velocity of outlet flow from this rotor occurs at the circumferential location of 25°. When a larger portion of the flow exits near the outlet cross-section, it travels a shorter length through the volute and experiences a smaller loss. Although it is theoretically expected that the flow should exit the rotor uniformly at all circumferential locations, this is generally not the case in squirrel-cage fans. Volute profile controls the flow through the rotor. A larger volute spread angle, and therefore a large volute cross-section would allow earlier maximum flow through the rotor. For the 5° volute spread angle, peak flow rate through the rotor occurs at latest possible location, which is 25° in this case.

Table 3.3 shows nondimensional integrals of the radial velocity for each traverse of the laser probe, and therefore the values indicate the flow through that section. This quantity is helpful since different profiles have local variations and are ambiguous for overall judgments. A notable trend is understood from a comparison of the radial velocity profiles for 270° and 360° in Figure 3.10 and Table 3.3. The ratio of radial velocities at two sections is reduced from E to A and then to D (Table 3.3). This confirms that the circumferential flow pattern is not the same for all cases, and it depends on the rotor geometry. The position of the peak outflow has corresponding results on efficiency. The reason for this behaviour is explained here and is developed later for other similar curves.

The flow that travels a larger path through the volute experiences a larger corresponding loss, and this results in smaller fan efficiency. In Figure 3.9(b) the efficiencies increase from E to A and then to D, and this is consistent with the trend

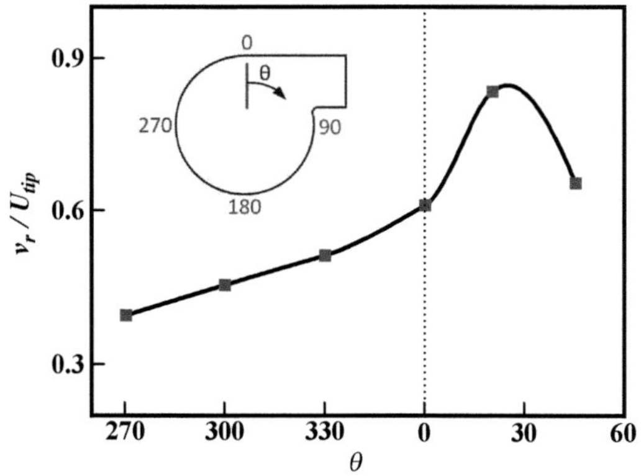

Figure 3.11 Radial velocity component in exit region of a $+10°$ half-cone rotor at various circumferential locations. Measurements are at $r/r_0 = 1.17$ and $z/B = 0.3$.

Table 3.3 A comparison of nondimensional radial outflow velocity components at two sections for three different cones at their respective maximum head, according to Figure 3.10

	θ	Operating point		
		A	D	E
Integral of normalized radial velocity component	270°	0.20	0.06	0.28
	360°	0.35	0.30	0.38
The ratio of integral of radial velocity at $\theta = 270°$ to that of $\theta = 360°$		0.57	0.20	0.74
Summation of two integrals		0.55	0.36	0.66

of the radial velocity components in Table 3.3. In other words, at point E there is comparable exit flow both at 270° and at 360° where the former travels a longer distance inside the volute, and therefore contributes to a smaller efficiency. The pattern changes as moving towards point A and then point D. A smaller share of flow goes through 270° and more flow moves towards later circumferential positions. Although flow rates are almost the same for the three points, the sum of the integrals for sections at 270° and 360° (Table 3.3) for point D is smaller than others. This means that other circumferential sections have stronger outflows.

The question of 'why the flow should exit an earlier or a later circumferential location' is a subject for future investigations, but one reason is the known fact that likewise flow in pumps, the volute of squirrel-cage fans also controls the flow through the rotor. A larger cross-section inside the volute at any location would induce more flow through the rotor there. At a small volute spread angle (<5°), the major part of flow goes through the rotor near the fan exit plane. A negative half-cone rotor as compared to other rotors has a relatively smaller space inside the volute close to the backplate, and therefore the flow exits the rotor as late as possible. A smaller volute width or more throttling would similarly move the circumferential location for maximum rotor exit flow.

Figure 3.12 presents more information on radial velocity trends for +10° and −10° conical rotors for operating points D and E, at $r/r_0 = 1.0$ and $r/r_0 = 1.17$. Two measurement sections at 270° and 360° are considered. In the case of +10° conical rotor (point D) at 270° (Figure 3.12(a)), the radial velocity decreases from $r/r_0 = 1.0$ to $r/r_0 = 1.17$. Since the flow at $r/r_0 = 1.17$ corresponds to exit flow from the rotor at some earlier circumferential locations, the rotor exit flow at 270° has increased as compared to previous circumferential locations. It is known that velocity reduction in the radial direction close to the rotor does not account for such changes. This means that for the present rotor and at this section, the flow out of the rotor is circumferentially increasing. As Figure 3.12(b) shows, an opposite trend is true for the same rotor at 360° and the outflow increases from $r/r_0 = 1.0$ to $r/r_0 = 1.17$. Accordingly, the maximum flow occurrs somewhere between the above

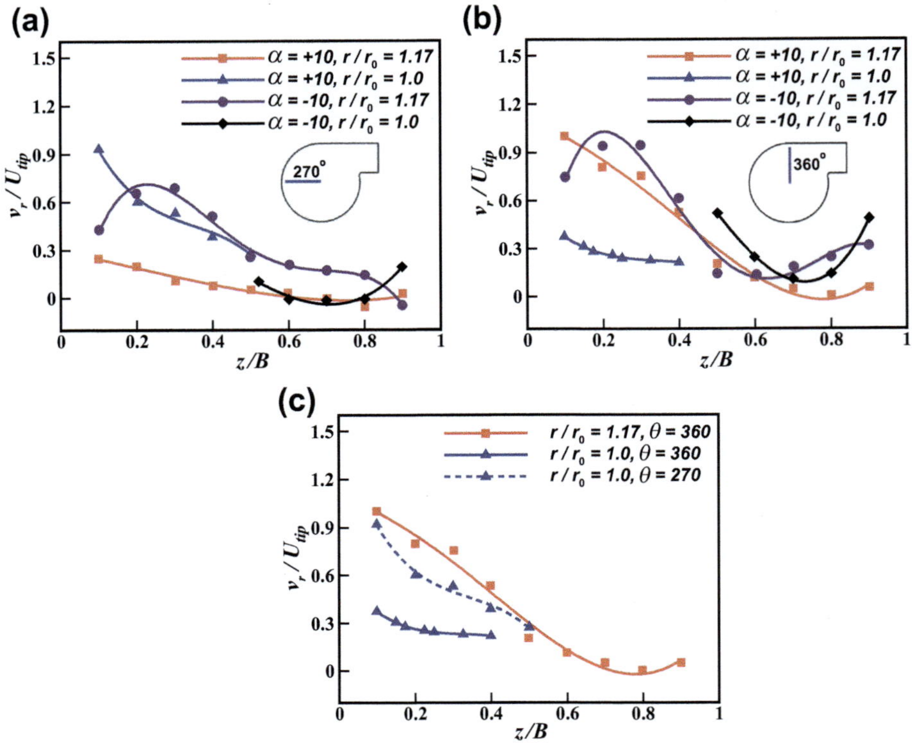

Figure 3.12 Radial component of normalized velocity in conical rotors for operating points D and E at $r/r_0 = 1.0$ and $r/r_0 = 1.17$. (a) Measurement at section $\theta = 270°$; (b) measurement at section $\theta = 360°$; and (c) repetition of curves for $+10°$ half-cone rotor and similarity of radial velocity profile for $r/r_0 = 1.0$ at $\theta = 270°$ and $r/r_0 = 1.17$ at $\theta = 360°$.

two sections. The outflow close to the cut-off is expected to reduce at such throttling, and the flow tends to circulate back into the rotor.

Conversely, for a $-10°$ half-cone rotor (point E) at 270°, radial velocity increases from $r/r_0 = 1.0$ to $r/r_0 = 1.17$. Therefore, there is more flow out of the rotor even before this section, which is decreasing, and therefore a less efficient fan is expected. Apparently, at this small flow rate, volute cross-section allows a uniform distribution of radial component in the circumferential direction, but as is shown later, this is changed at larger flow rates. Table 3.3 shows that the $-10°$ half-cone rotor has comparable rotor outflow at 270° and 360°. A large part of the rotor circumference therefore experiences outflow. Figure 3.12(b) for 360° section shows little change with radius and behind the inlet for the negative half-cone rotor. The smaller traverse for this rotor is located just behind the inlet (inside the separated flow) where pressure gradients and three-dimensional effects inside the volute also control the velocity profiles.

Figure 3.12(c) compares the radial velocity profile of the $+10°$ half-cone rotor for $r/r_0 = 1.0$ at $\theta = 270°$ with that of $r/r_0 = 1.17$ at $\theta = 360°$. A similar profile confirms

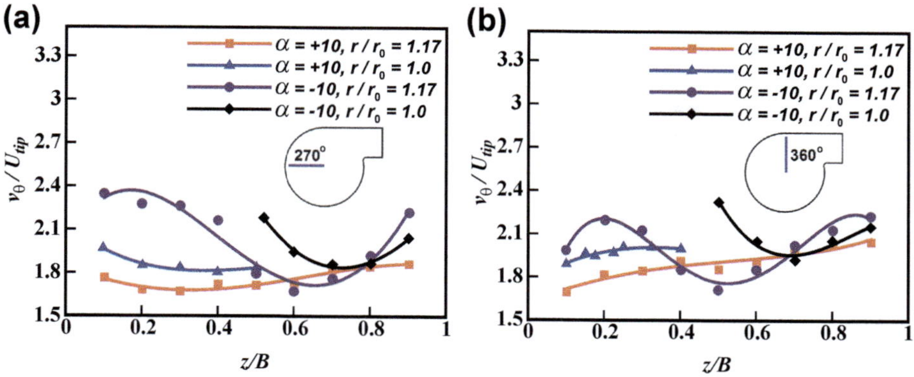

Figure 3.13 Circumferential component of normalized velocity in the conical rotors at $r/r_0 = 1.0$ and $r/r_0 = 1.17$. Measurements are at sections (a) $\theta = 270°$ and (b) $\theta = 360°$.

that the flow at the larger radius ($r/r_0 = 1.17$) at $\theta = 360°$ could have come out of the rotor at previous locations close to $\theta = 270°$.

Figure 3.13 shows the circumferential velocity profile at $r/r_0 = 1.0$ and $r/r_0 = 1.17$ for sections at 270° and 360°. These graphs show that the circumferential velocity marginally increases with radius but does not show any trends for a change in circumferential location.

3.1.3.2 Larger head or a more efficient fan?

Points B and C in the performance curves of Figure 3.9 correspond to the cylindrical and the +5° half-cone rotors, respectively, with a similar head coefficient and different efficiency (C is more efficient). At similar flow coefficients, B on the cylindrical and F on the −10° conical rotor curves have a similar efficiency, but there is a higher head for point F. This shows that within the usual operating range of fans, the positive cone rotor produces a more efficient fan with no loss of head and the negative cone rotor can produce higher heads with no loss in efficiency. The cone angles for the considered rotors are selected with equal angular spacing, but the results show that there are no similar optimum angles for positive and negative cones. This section relates the added efficiency to the circumferential distribution of the radial velocity, and in this way reconfirms the previous conclusions. It is shown later that different lean angles are also responsible for both velocity patterns and efficiencies.

Figure 3.14 shows the velocity profiles of cylindrical (point B), +5° half-cone (point C) and −10° half-cone (point F) rotors at sections 270° and 360°. The surface area underneath the radial velocity curves is presented in Table 3.4 and provides a basis for comparison between the local flow rates through different sections of the rotor.

Similar values in the last row of Table 3.4 comply with the similar fan flow rates for operating points B, C and F. This means that the flows through these two sections are indicative of the fan overall flow rate. For the cylindrical rotor (point B), the exit flow rate at section 270° is 1.82 times that of section 360°. In contrast, for a +5° half-cone rotor (point C), the flow which exits the rotor at section 270° is only about 0.57 times

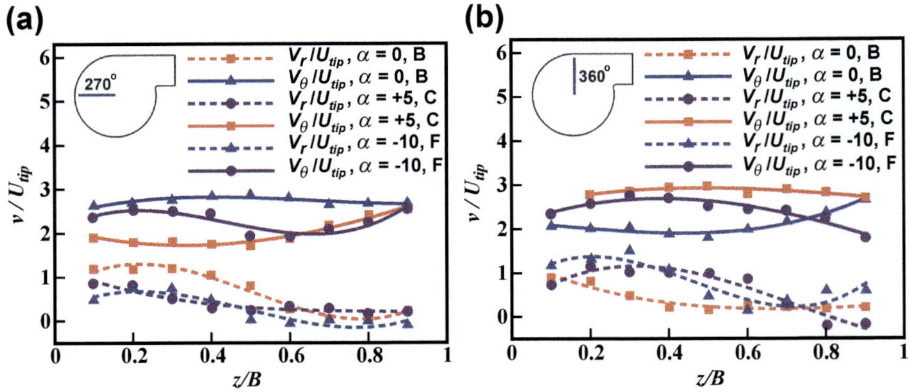

Figure 3.14 Axial variation of normalized velocity components (radial and circumferential) at $r/r_0 = 1.17$ and $\Phi = 0.55$ for three different rotors. Letters B, C and F in the legend indicate the operating condition points. Measurements are at sections: (a) $\theta = 270°$ and (b) $\theta = 360°$.

Table 3.4 A comparison of nondimensional radial outflow velocity components at two sections for operating points B, C and F, according to Figure 3.9

		Operating point		
	θ	B	C	F
Integral of normalized radial velocity component	270°	0.53	0.31	0.21
	360°	0.29	0.54	0.64
The ratio of integral of radial velocity at $\theta = 270°$ to that of $\theta = 360°$		1.82	0.57	0.33
Summation of two integrals		0.82	0.85	0.85

that of section 360°. Therefore, the reason for a higher efficiency in the +5° rotor is that a larger part of rotor exit flow occurs at 360°. This flow has a shorter path to travel inside the volute, and a larger efficiency would be therefore ensued.

Point F does not seem to comply with this trend at the first sight and even for a smaller velocity ratio between the two sections it is just as efficient as point B. Of course, fan efficiency is not only a function of circumferential distribution of velocity profiles, and this −10° rotor is notably inefficient at smaller flow rates (point E). A comparison between points F and E (that locate on the same performance curve) is shown in the next section (Table 3.6): the added flow rate at point F occurs at 360° with a flow reduction at 270°. Although the fan is less efficient at smaller flow rates, added flow at 360° brings the efficiency at operating range to comparable values with the cylindrical rotor. A higher head for point F was explained previously, and it is because of a larger blade radius and the work done on the flow through the rotor.

Table 3.5 A comparison of nondimensional radial outflow velocity components at two sections for operating points A and B, according to Figure 3.9

	θ	Operating point	
		A	B
Integral of normalized radial velocity component	270°	0.20	0.53
	360°	0.35	0.29
The ratio of integral of radial velocity at $\theta = 270°$ to that of $\theta = 360°$		0.57	1.82
Summation of two integrals		0.55	0.82

Table 3.6 Comparison of nondimensional radial outflow velocity components at two sections for operating points E and F, according to Figure 3.9

	θ	Operating point	
		E	F
Integral of normalized radial velocity component	270°	0.28	0.21
	360°	0.38	0.64
The ratio of integral of radial velocity at $\theta = 270°$ to that of $\theta = 360°$		0.74	0.33
Summation of two integrals		0.66	0.85

3.1.4 Influence of flow rate on efficiency and velocity components

Figure 3.15 shows the velocity profiles in the cylindrical rotor for different flow rates at points A and B. This figure and Table 3.5 (that compares the radial outflow at sections $\theta = 270°$ and $\theta = 360°$) show that the rotor exit flow at section $\theta = 270°$ increases with the flow rate. The table shows that the ratio of outflow between the two sections increases from 0.57 to 1.82, as the flow rate increases. At point B, larger volume of the flow travels a longer distance inside the volute, and a corresponding decrease in efficiency is ensued. Because this is a forward-curved rotor, for an increase in the radial component, a corresponding increase in the tangential component is justified. Reduction of fan throttling activates more locations around the rotor to pass the added flow. If, like the present case, this happens in an earlier circumferential location, an efficiency reduction could be ensued. If, like the next case, it occurs in later

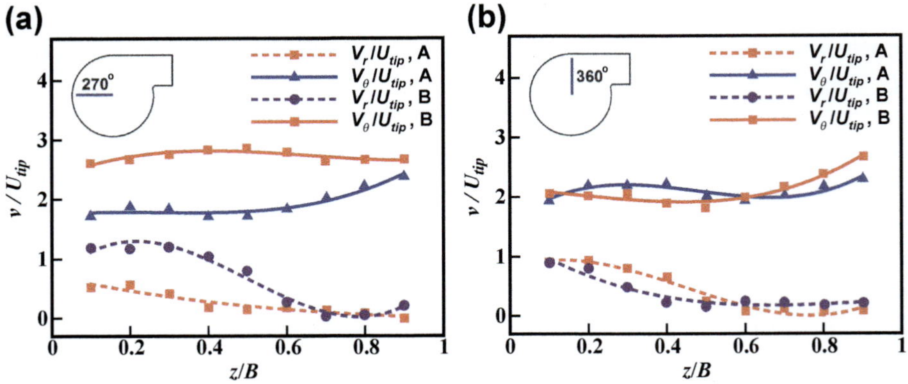

Figure 3.15 Axial variation of normalized velocity components (radial and circumferential) at two different operating conditions for cylindrical rotor. Measurements are at (a) $\theta = 270°$ and (b) $\theta = 360°$.

circumferential locations in the volute, then the added flow would travel a shorter distance within the volute, and this would result in a better efficiency.

Figure 3.16 and Table 3.6 present the velocity profiles and rotor outflow for the $-10°$ half-cone rotor at two different flow rates (points E and F). The figure shows that the added flow rate due to opening of the throttle causes the flow to exit from the section at 360°. A better efficiency for point F as compared to point E therefore ensues. This complies with other results that the volute could block the flow out of the rotor through a smaller volute cross-section. The negative rotor provides this smaller volute cross-section at locations closer to the backplate where the main part of the outward flow from the rotor occurs. This pushes any extra flow towards later circumferential locations and enhances efficiency.

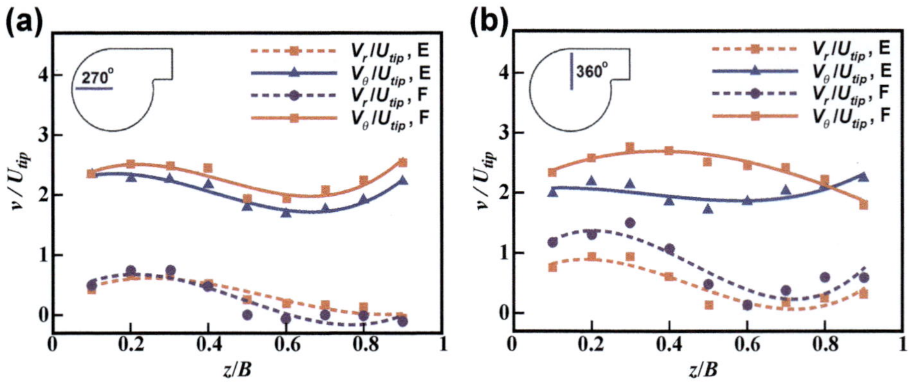

Figure 3.16 Axial variation of normalized velocity components (radial and circumferential) at two different operating conditions for the $-10°$ half-cone rotor. Measurements are at (a) $\theta = 270°$ and (b) $\theta = 360°$.

3.2 Lean angle in rotor blades

Although the discussion in Section 3.1 sheds some light on the fluid mechanics of half-cone rotor properties, it still does not explain the optimum cone angles since fan loading and efficiency have only opposite trends.

Leaning the blades is another modification that affects the performance of squirrel-cage fans. It is angular alignment of the blades relative to the rotor axis so that the blade edge and the axis are no longer coplanar. The blade arc radius and length are similar for the purpose of comparison for both the cylindrical and half-cone rotors whether the blades are leaned or not. The lean angle (β) is defined to be positive when it is in the direction of rotation and if otherwise, it is considered as a negative lean angle (Figure 3.17).

3.2.1 *Modelling for the optimum lean angle*

Numerical simulation of fluid flow in the fan with half-cone rotor shows that efficiency steadily increases with cone angle, and there is no optimum at $\alpha = +5°$. Simulation of the effect of leaning the blades for cylindrical and half-cone rotors shows no improvement for cylindrical and negative half-cone rotors. For positive half-cone rotors, two specific lean angles of $\beta = \pm 2°$ would increase both the head coefficient and efficiency (Figure 3.18). This is a trend that its real value could only be verified with experiments.

The next section is an experimental clarification of individual and combined effects of a $+10°$ half-cone rotor and a $-2°$ alignment of blade. The experiments include performance tests and velocity measurements with a laser Doppler anemometer. They are then analysed together to show the reasoning behind the improvements.

3.2.2 *The optimum lean angle*

When one compares squirrel-cage fans of different geometries, a decision on appropriate reference dimension is inevitable. It is common in squirrel-cage fans, similar

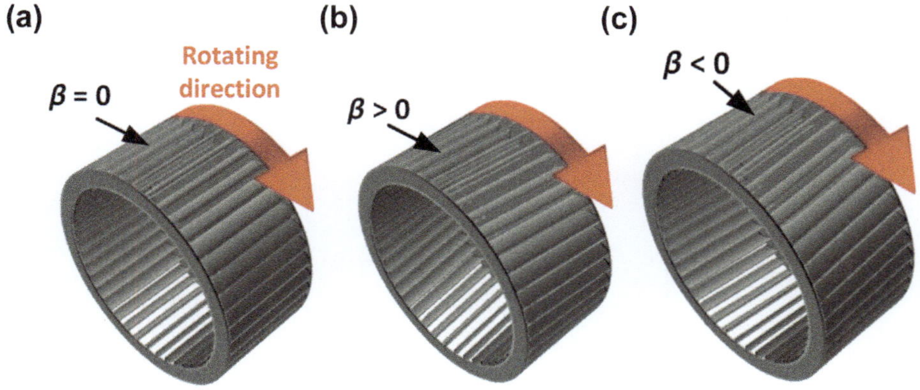

Figure 3.17 Cylindrical rotors with different blade alignment: (a) a rotor without lean angle; (b) a rotor with positive lean angle; and (c) a rotor with negative lean angle.

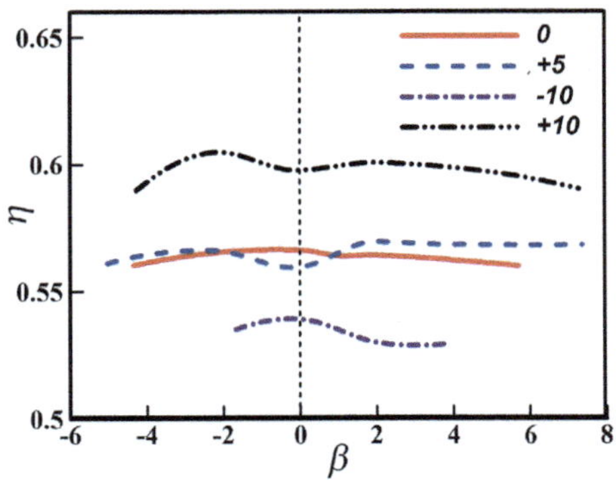

Figure 3.18 Prediction of change in efficiency with lean angle of blades for cylindrical and half-cone rotors.

to any other turbomachine, to select rotor diameter as reference length. An average length that represents volute dimensions would also be appropriate, since the important dimension for the customer is the volute dimension and not the rotor diameter. It is shown in Chapter 4 that it is possible to use a length scale representing overall fan dimensions and in this method, a volute with 5° of spread angle produces the maximum head for that fan size. It is therefore possible to select any size of the half-cone rotor inside this space as long as volute dimensions are fixed.

Half-cone and cylindrical rotors were compared in the previous section by selecting rotor mid-width diameter of the half-cone rotor as the reference. A comparison of the radial component of outlet velocity in different rotors shows that the maximum radial velocity exits at the axial distance of $0.2 < z/B < 0.4$ from the backplate. Since a large portion of flow goes through the rotor there, a reference diameter at $z/B = 0.3$ would be more appropriate in an extension to the above discussion. This would provide equal peripheral speeds between cylindrical and half-cone rotors, not at the geometric location of the rotor mid-width, but where maximum energy is transferred to the flow and therefore similar head coefficients are expected (Figure 3.19). This might pose some consideration about generated noise, but since the cut-off distance for the position where most of the rotor outflow occurs is not changed, a significant variation in sound level is not expected.

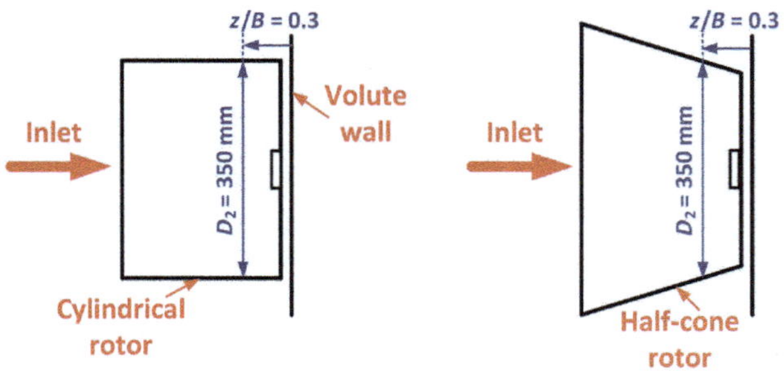

Figure 3.19 Reference diameter for a cylindrical rotor and its corresponding half-cone rotor.

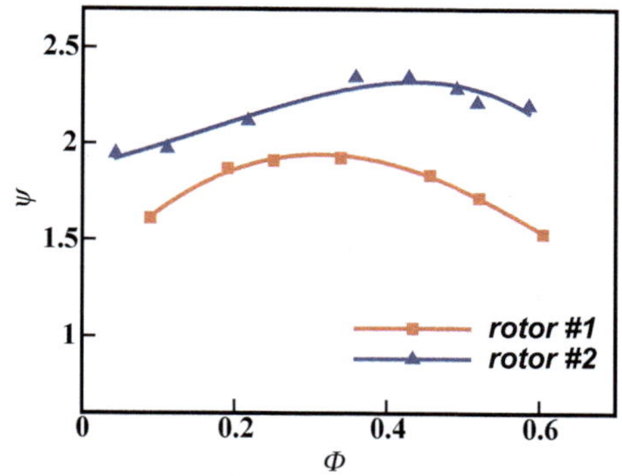

Figure 3.20 Total head coefficient of +10° half-cone rotors with different reference diameters: rotor #1 with the average diameter as the reference diameter and rotor #2 with the diameter at $z/B = 0.3$ as the reference.

The performance curve for the +10° half-cone rotor with two different reference diameters is shown in Figure 3.20. In rotor #1, the average rotor diameter at $z/B = 0.5$ is considered as the reference diameter, while the reference in rotor #2 is considered to be the diameter at $z/B = 0.3$. The rotor #2 is a larger rotor in a similar casing and as is expected, results in higher head coefficients.

Individual and combined effects of rotor cone angle and blade lean angle are examined by comparing the performance of three different rotors, namely, a cylindrical rotor, a +10° half-cone rotor with no lean angle and a +10° half-cone rotor with −2° lean angle (Figure 3.21). All other fan dimensions are the same and selected from optimized values on the basis of previous studies. Only the inlet bell-mouth diameter is changed for each rotor to match with its inlet diameter. According to Figure 3.21 both +10° half-cone rotors have improved or at least similar head coefficients occur at the right region of maximum head point where most fans operate. Also at the maximum flow coefficient, both half-cone rotors have approximately similar head coefficients.

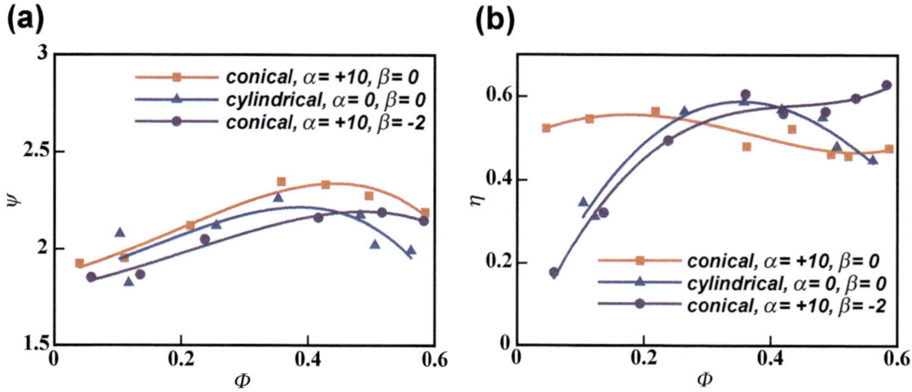

Figure 3.21 (a) Total head coefficient and (b) efficiency of fans with cylindrical rotor, +10° half-cone rotor without blade lean angle and +10° half-cone rotor with −2° blade lean angle.

Figure 3.21(b) shows that the efficiency of the $+10°$ half-cone rotor with no lean angle is higher than the cylindrical rotor in low and also maximum flow coefficients. The main attractive point of this rotor is its approximately constant efficiency at different flow coefficients. Nowadays fan manufacturers use control technology (such as speed regulation) for producing fans with approximately uniform characteristics at different flow rates. It is even more desirable to design such a fan but with no extra equipment. When blades are leaned, the efficiency curve rises continuously and has higher efficiencies than both rotors at average to high flow coefficients.

The $+10°$ half-cone rotor with $-2°$ leaned blades as compared with the cylindrical rotor indicates about 0.2 of enhancement in the head coefficient and about 18.3% increase in the efficiency at the maximum flow coefficient.

3.2.3 Velocity profiles

Laser velocimetry procedure in this section determines the velocity profiles in order to trace vortex location behind the inlet and mass flow distribution through the rotor.

In order to compare the flow field in the half-cone rotors with and without lean angle, the velocity components of different axial locations (Figure 3.22) at an angular section of 25° and at $r/r_0 = 1.17$ are measured. This radial position is chosen to avoid blockage of laser beams by rotors. The change in velocity profiles from rotor outlet to this radius is negligible.

Figure 3.23 compares radial and circumferential velocity components at different flow coefficients (0.52 and 0.43) for two $+10°$ half-cone rotors, one without lean angle and the other with a $-2°$ lean angle. Little if any difference is evident between velocity components of the two rotors for either flow coefficient. Radial velocity of outflow from a rotor is an indication of the fan flow rate. Hence, similarity of radial velocity

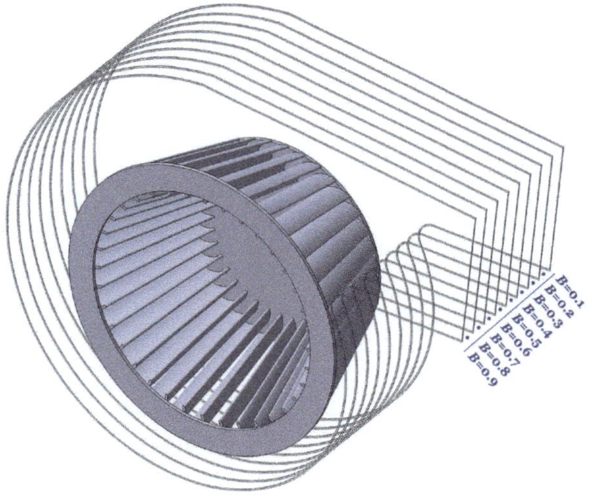

Figure 3.22 The axial sections in the volute in which the velocity components are measured.

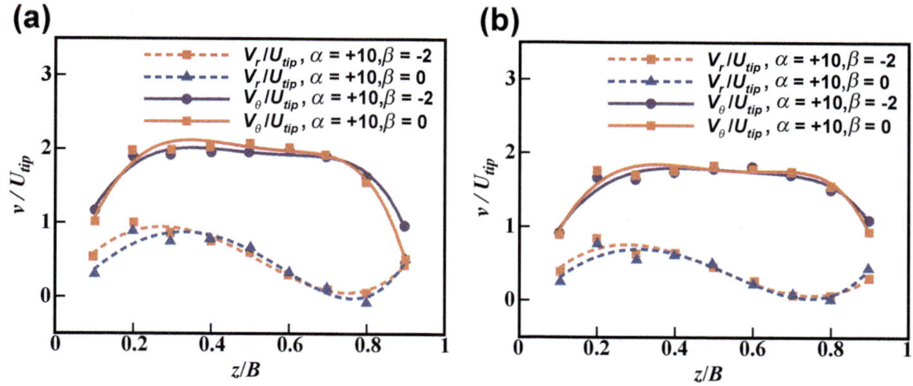

Figure 3.23 Axial variation of radial and circumferential velocity components in two fans with $+10°$ half-cone rotors, one without leaned blades and the other with $-2°$ leaned blades. Measurements are at different flow coefficients: (a) $\Phi = 0.52$ and (b) $\Phi = 0.43$.

in $+10°$ half-cone rotors with and without leaned blades is consistent with equality of their flow coefficient. Also similarity of circumferential velocity component at outlet of these two rotors expresses the equality of transmitted power to the fluid. This is consistent with the similarity of head coefficients for both fans at high but equal flow coefficients.

Although radial and circumferential velocity components show similar trends for the two rotors, axial velocity components are different. Before presenting the axial profiles, Figure 3.24(a) repeats the expected vortex pattern behind the inlet of these fans. This clarifies that the axial velocity (z component) is a measure of vortex location behind the inlet. Figure 3.24(b) shows the change of the axial velocity with radius along $z/B = 0.7$, where the vortex centre is expected to move. Positive axial component at lower radius (close to the rotor exit) shows that the flow has

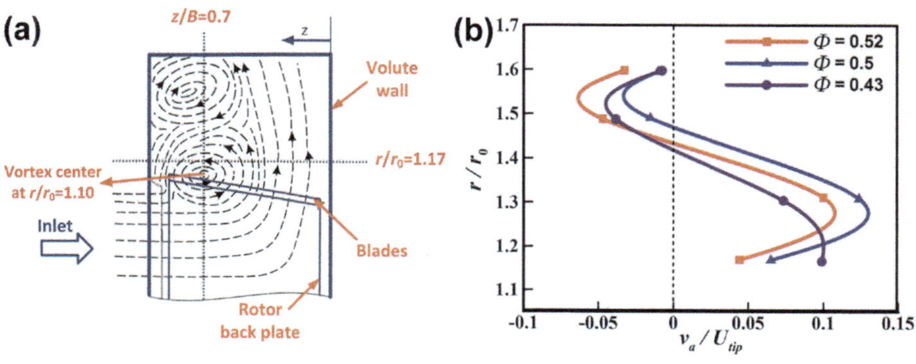

Figure 3.24 (a) Visual representation of the expected vortex behind the fan inlet and (b) variation of axial velocity component with radius along $z/B = 0.7$ for a $+10°$ half-cone rotor with no lean angle and for various flow coefficients.

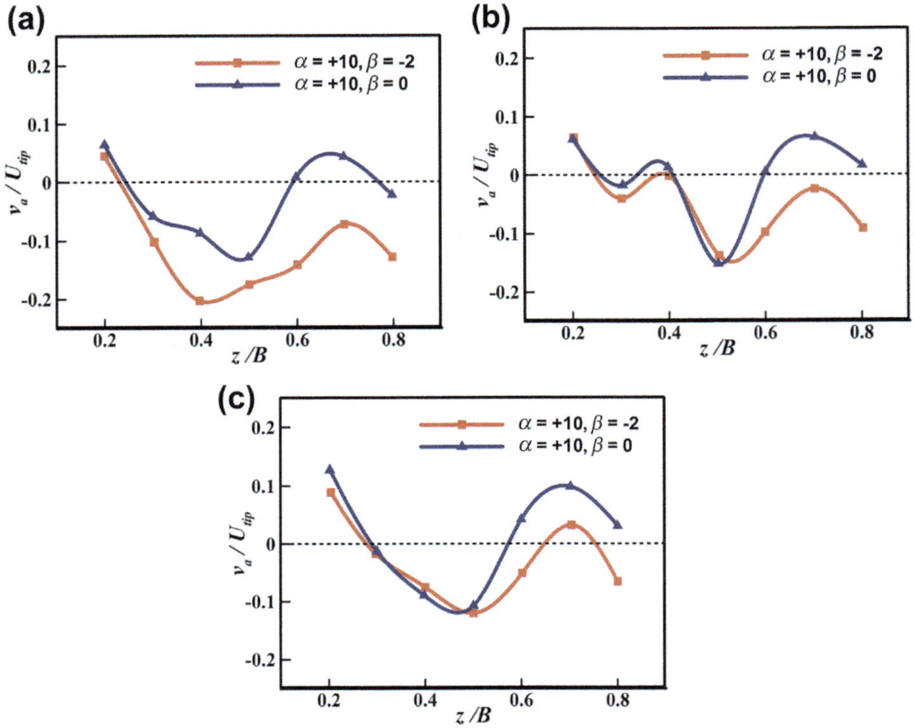

Figure 3.25 Axial velocity profile in two $+10°$ half-cone rotors, one without leaned blades and the other with a $-2°$ lean angle. Measurements are at various flow coefficients: (a) $\Phi = 0.52$; (b) $\Phi = 0.5$; and (c) $\Phi = 0.43$.

turned towards the inlet. Negative axial component at higher radius (far from the rotor exit) belongs to another vortex farther away from the rotor. The centre of the vortex behind the inlet is expected to be along $r/r_0 = 1.1$ from an extrapolation of data in Figure 3.24(b). When blades have no lean angle, there is return flow behind the inlet at all flow coefficients.

Figure 3.25 compares the axial velocity profile for different flow coefficients (0.52, 0.5 and 0.43) for the two $+10°$ half-cone rotors, with and without leaned blades. This is an indication of return flow after the inlet at $r/r_0 = 1.17$. In the case of the half-cone rotor with no lean angle almost half of the rotor width behind the inlet ($0.55 < z/B < 1.0$) experiences positive axial velocities (flow towards the inlet). When blades are leaned, negative axial component occupies a larger portion of the rotor. This means that the vortex behind the inlet has moved to a larger radius. A marginal respective increase in radial component is also evident in Figure 3.23 that confirms the vortex movement. At higher flow rates for the rotor with leaned blades, the vortex has totally moved above the rotor. Positive radial and axial components for the half-cone rotor with leaned blades indicate that the flow turns immediately after the inlet and passes through all rotor width which removes the blade inactive zone.

Further reading

Alinejad, J., Montazerin, N., Damangir, A., & Samarbakhsh, S. (2007). Experimental and numerical analysis of increasing efficiency in squirrel cage fan with half cone rotors. in Persian. In *Proceedings of the 15th international conference on mechanical Engineering (ISME 2007)* (pp. 587–589). Tehran, Iran: Amirkabir University of Technology.

Damangir, A. (2004). *Joint impeller/scroll sizing of squirrel cage fans using alternative non-dimensional head and flow rate coefficients* (Ph.D. thesis). Amirkabir University of Technology.

Eck, B. (1973). *Fans* (1st ed.). Oxford: Pergamon Press (pp. 117–40).

Kind, R. J., & Tobin, M. J. (1990). Flow in a centrifugal fan of the squirrel cage type. *ASME Journal of Turbomachinery, 112*, 84–90.

Morinushi, K. (1987). The influence of geometric parameters on FC centrifugal fan noise. *Journal of Vibration and Acoustics Stress and Reliability in Design, 109*, 227–233.

Nikkhoo, M. (2009). *Investigation on fluid flow inside squirrel-cage fan with frustums or right circular cones rotor* (in Persian) (MSc. thesis). Amirkabir University of Technology.

Nikkhoo, M., Montazerin, N., Damangir, A., & Samian, R. S. (2009). Analysis of the effect of blade torsion in cylindrical and half cone rotors in squirrel cage fan performance. in Persian. In *Proceedings of the 17th international conference on mechanical engineering (ISME 2009)* (pp. 1039–1041). Tehran, Iran: University of Tehran.

Rezaei Niya, S. M., Montazerin, N., Damangir, A., & Dehkordi, A. H. (2006). Performance and laser Doppler anemometry experimental investigation of squirrel cage fans with half-cone rotors. *Proceedings of the Institution of Mechanical Engineers, Part A: Journal of Power and Energy, 220*, 753–763.

Roth, H. W. (1981). Optimierung von trommelläufer-ventilatoren. *Strömungsmechanik und Strömungsmaschinen, 29*, 1–45.

Velarde-Suarez, S., Ballasteros-Tajadura, R., Santolaria-Morros, C., & Gonzalez-Perez, J. (2006). Unsteady flow pattern characteristics fans. *Journal of Fluids Engineering, 220*, 753–763.

Volute

4

4.1 Volute flow

The volute of a centrifugal fan collects the flow from blade cascades and delivers it to the exit. The pressure recovery takes place in a radial diffuser which is either located between the rotor and the volute or in a diffuser just after the throat. The structure of the flow in this region is complicated and involves separated, reverse and vortical flow zones.

Volute action in squirrel-cage fans is even more pronounced than what is generally known for centrifugal turbomachines. Tests have shown that if the volute is removed, only little flow goes through the rotor inlet at all.

When there is a volute, proximity of the volute to the rotor acts like a valve on the rotor. A small volute spread angle limits the flow but results in a more aerodynamic flow through the blades. A large volute spread angle for the same rotor acts far better than any other modification. The flow rate could be even doubled for such a configuration, but causes a less aerodynamic flow.

4.1.1 Rotor–volute interface

Flow structures in the rotor exit region of forward-curved fans are generally three dimensional and unsteady. These complexities originate from intensive interactions between phase-dependent flow structures and moving/stationary parts of the fan. A few consequences of these interactions are transient wakes, flow separation, significant flow turns, acoustic noise, vibration of various components and flow nonuniformities. They significantly affect efficiency and performance of a turbomachine. Detailed investigations of flow field in the rotor exit region plus modifications in the geometry and/or operating condition further clarify these deficiencies.

The volute has a variable cross-section that collects jets and wakes, mixes them with the flow from previous circumferential locations and finally delivers the fluid to the outlet. Multiblade forward-curved centrifugal fans have a larger rotor exit-to-inlet area ratio and a larger number of rotor blades in comparison with centrifugal pumps and compressors. Flow structures after the rotor and at volute inlet are consequently more three dimensional, and the interaction between jet–wake structures can be stronger due to a smaller blade-to-blade distance.

The volute flow is even more complex at off-design where the flow partially returns back into the rotor. This gives stalled recirculation zones within the blade passages and creates backflow along the blade pressure side. This extension of separated flow into the volute complicates the flow field near cut-off. The nonuniformity of discharge flow from the impeller is interpreted as a result of secondary flow extension in turbomachine.

Developments in Turbomachinery Flow.

Flow field at the outside zone of the rotor is significantly more nonuniform in comparison with rotor inside zone. Still, flow separation and wakes at the blade suction sides as well as the shroud-induced separation are the most important reasons for complex flow patterns at the rotor exit region.

4.1.2 Jet–wake structures

The flow field inside the volute and in the rotor exit region of centrifugal fans starts with jets and wakes and continues to their interactions with the volute flow. The large number of blades (shorter blade-to-blade distance) and the large width-to-diameter ratio of the rotor intensify the interaction between jet and wake structures.

The interactions among jet and wake structures, rotor and stator blades and cyclic flow through the casing are common phenomena in various turbomachines. Generated wakes by rotor and stator blades are characterized by high-turbulent kinetic energy, elevated deformations and vortical flow structures.

Circumferential and axial variations in flow characteristics of the rotor exit region are due to asymmetric geometry of volute and one-sided entrance of fluid into the squirrel-cage fan. The flow field downstream of the rotor blades contains signatures from the inlet velocity profile, turning flow into the rotor and flow through the blades.

Figure 4.1, depicts sample normalized radial velocity scatters for four radial positions in the middle of the rotor width ($z/B = 0.5$) where through-flow is present.

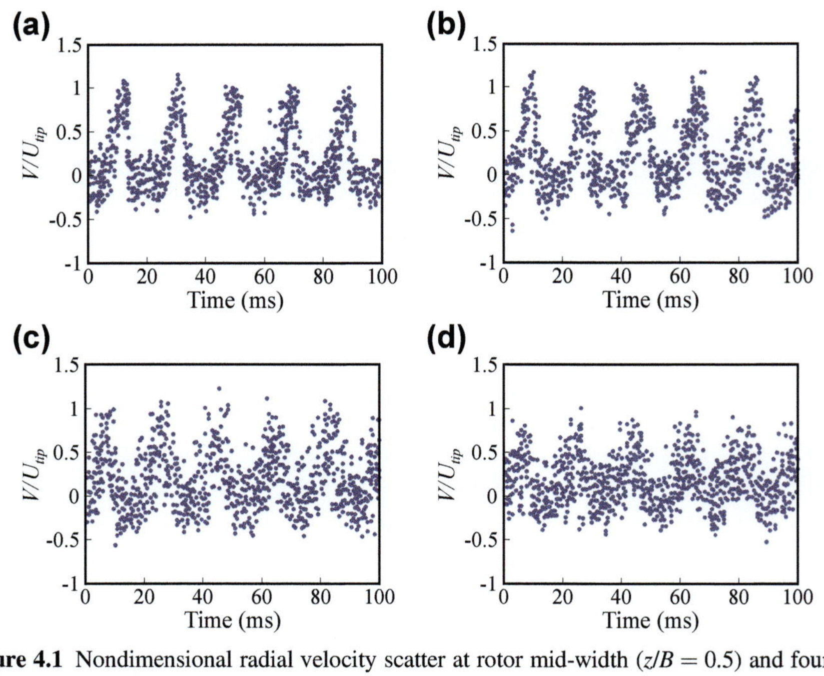

Figure 4.1 Nondimensional radial velocity scatter at rotor mid-width ($z/B = 0.5$) and four nondimensional radial positions: (a) $r/r_0 = 1.03$; (b) $r/r_0 = 1.09$; (c) $r/r_0 = 1.15$; and (d) $r/r_0 = 1.21$.

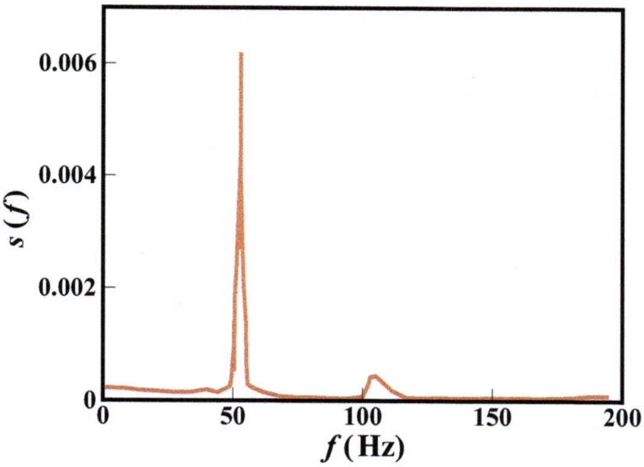

Figure 4.2 A sample spectrum of the radial velocity for a point close to the rotor exit. The peak at 53 Hz corresponds to blade passing frequency.

The captured time corresponds to approximately six passing blades in front of the measurement volume. The diverse range of measured velocities is due to the blade passing effect, which is also the origin of this unsteady turbulent flow. The presence of jet—wake pattern as the typical characteristic for radial fans and pumps is evident in Figure 4.1(a). In this case, the wake has a major share, which is a symptom of the flow separation on the blade suction side. Figure 4.1(b)—(d) shows that turbulent jets originated from the rotor mix with blade wakes as the radius increases. It becomes therefore more difficult to distinguish jet and wake regions from each other as the radius increases.

Figure 4.2 presents the power spectral density function of the radial velocity at a sample point after the rotor. The spectrum is dominated by a peak at the blade passing frequency (53 Hz in this experiment) and its harmonic (106 Hz). This means that the velocity profile is mainly affected by rotor-induced jet—wake patterns.

Figure 4.3 presents the averaged radial velocity and its root mean square along the rotor width at different radial positions. Outward radial velocity covers almost half of

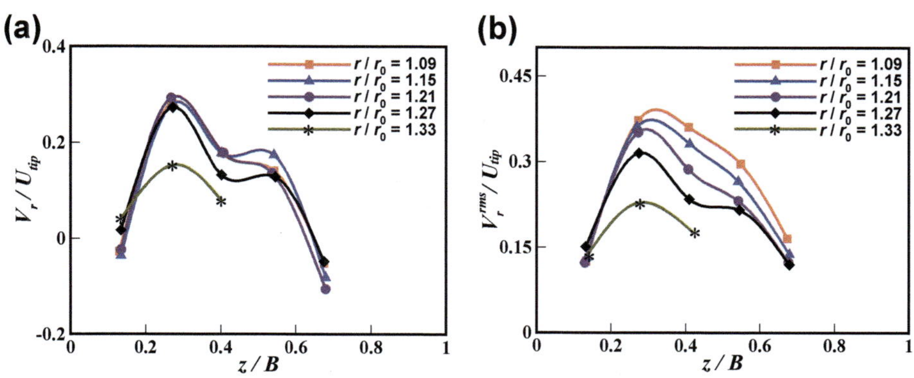

Figure 4.3 Nondimensional radial velocity profile (a) and its root mean square (b) along the rotor width at different radial stations.

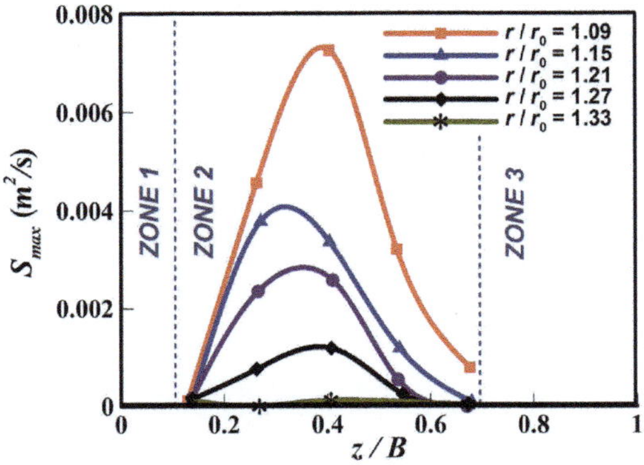

Figure 4.4 Power spectral density of the radial velocity at the frequency of 53 Hz in the meridional plane at different radii.

the volute width and the flow turns towards the rotor near the sidewalls (particularly close to the inlet-side). Root mean square of radial velocity has a maximum in the middle of rotor width which does not diminish near the walls. The dominant feature is that this root mean square exceeds the averaged radial velocity.

Axial variation of power spectral density at a frequency of 53 Hz is investigated in Figure 4.4 for different radial positions in the section $\theta = 270°$. According to Figures 4.3 and 4.4, the volute width may be divided into three different zones:

1. First is the distance between the volute wall and the rotor backplate ($z/B < 0.1$), which occupies about 13% of the volute volume in this case. The average velocity is small at this zone and velocity fluctuations are in the order of 0.1 m/s. Figure 4.4 does not show any dominant spectral density for this zone which means that blade passing action is absent.
2. The second zone is the distance from the rotor backplate to almost 60% of the volute width, where average radial velocity at the rotor exit and its root mean square are significant. This means that most of the flow passes through this zone. Blade passing frequency is present and the strength of the spectral density function at this frequency reduces as the radius increases (Figure 4.4).
3. Stalls happen in locations closer to the inlet (zone 3, $z/B > 0.7$) due to flow separation from the inlet lip. Behind the inlet, only a small area of the rotor periphery senses radial velocity out of the rotor, and a larger part experiences no flow. The blade passing frequency is absent in any spectrum at this separating zone. It is believed that this active stall cell behind the inlet lip interacts with the main flow and rotates even faster than the rotor. Therefore, this area may not be a source of noise with blade passing frequency.

4.1.3 Circumferential distribution of jet and wake structures

When the fan flow rate is low, larger portions of flow exit at earlier angles after the cut-off. For larger flow rates, bulk of the exit flow from the rotor moves to larger volute cross-sections, closer to the fan exit. It is mentioned earlier that the volute controls the

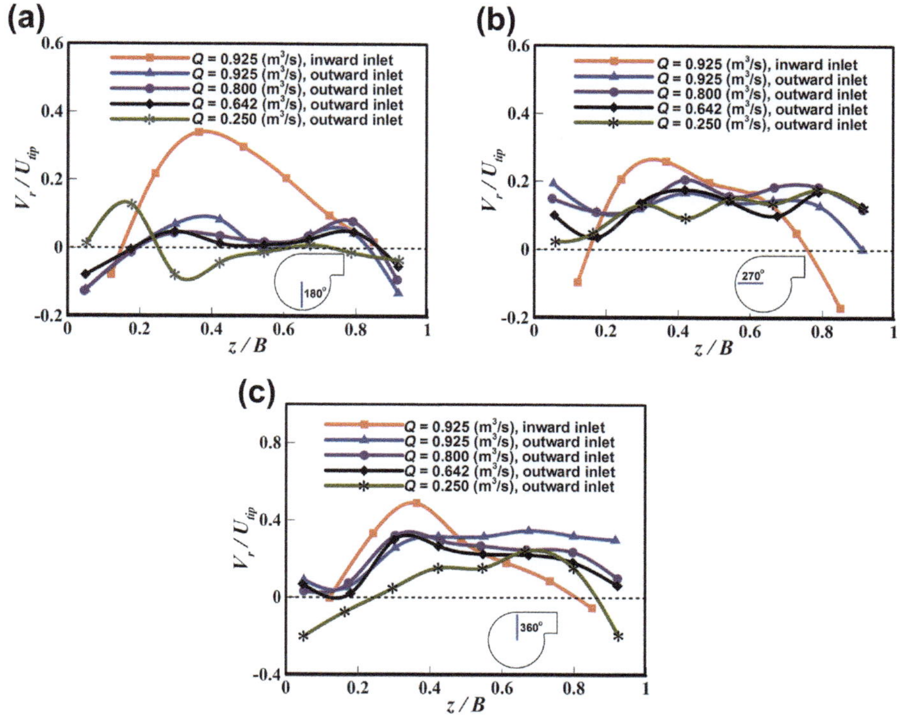

Figure 4.5 Normalized radial velocity component for different inlet configurations and various flow rates at three circumferential sections: (a) $\theta = 180°$; (b) $\theta = 270°$; and (c) $\theta = 360°$.

flow through the rotor. This is also true that the largest portion of flow selects a volute angle with an appropriate cross-sectional area relative to the total flow rate.

Figure 4.5 illustrates the nondimensional radial component of velocity at three circumferential sections (at 180°, 270° and 360°). Measurements are for both inward and outward inlets and for various channel throttling (different flow rates). Two major points can be interpreted from Figure 4.5:

1. The radial velocity component and consequently the flow rate increase as the volute cross-sectional area increases. The jet–wake–volute interactions are therefore intensified in that direction, and the strongest interactions occur around the section located at 360° (near the volute outlet).
2. The nonuniform velocity profile for the inward inlet configuration is replaced with fairly uniform velocity in the case of outward inlet configuration.

4.2 Slip factor

It is known that the flow out of the impeller of centrifugal turbomachines does not fully follow the blade exit angle. The flow separates from blade suction surfaces long before the blade trailing edge. Large variations in flow direction can occur from blade to

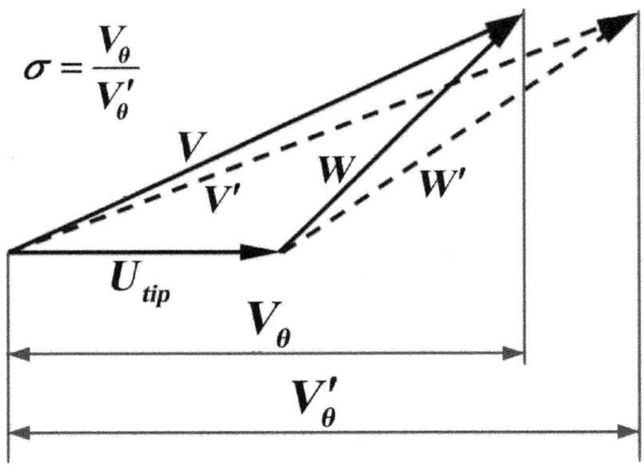

Figure 4.6 A definition for slip factor from ideal (broken lines) and real (solid lines) velocity triangles at the rotor outlet. U_{tip} is the rotor peripheral speed, V is the fluid velocity vector and W is the relative velocity vector.

blade, especially in the cut-off neighbourhood. This and other viscous effects exert a large flow deviation, and therefore the blade angle plays a less significant role in guiding the flow. Slip factor that contributes in the Euler energy equation is widely used to show this deviation. It is similar to the familiar 'work done factor' and helps the designer to predict the performance at an early stage of the design process.

Slip factor is defined as the ratio of real tangential velocity out of the rotor to its ideal value (Figure 4.6). It is calculated both theoretically and experimentally where the latter is obtained either directly from velocity profile or indirectly from flow rate and torque measurements.

The above definition is valid for all centrifugal turbomachines, but in squirrel-cage fans velocity profiles are strongly nonuniform. Volumetric flow rate and torque do not adequately represent the velocity components out of the rotor. An overall slip factor could therefore not represent the true deviation of the local tangential velocity from its ideal value.

Ideal flow predictions use blade geometry and flow parameters to calculate the relative eddy and slip factor of centrifugal turbomachines. The relationship for slip factor that is based on calculation of eddy inside the blade passage is acceptable for long backward-curved blade passages and a smaller number of blades than that of this fan. Extensions of this method acceptably predict the effect of the finite number of blades and flow coefficient for backward-curved fans. It is not accurate for power prediction in large curvature blades of forward passages. This is also due to a large rotor diameter ratio, highly forward-curved blades and the very nonuniform flow fields.

4.2.1 Experimental measurement of slip factor

The available slip factor correlations have not helped the design of squirrel-cage fans. If the flow does not follow the blades (and it is not even expected to do so), then how do geometric parameters control the flow field in this fan? The common approach in

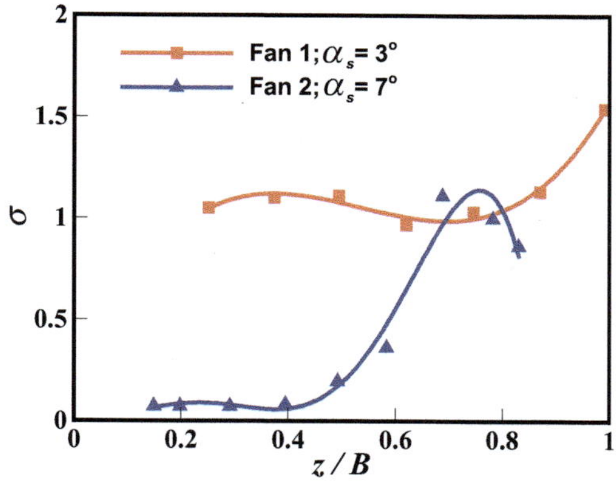

Figure 4.7 Axial variation of slip factor for two different fans. Measurements are at maximum flow rate and $\theta = 360°$.

this case is that a set of general design rules are outlined, and the performance data are then experimentally obtained after manufacturing the fan. Simulation of the flow in squirrel-cage fans has resulted in limited success for such purpose.

Deviation of flow from the blade orientation is more complicated here than in other centrifugal turbomachines. The tangential component of velocity is not the only parameter that changes the slip factor. Combination of a large radial velocity and a large blade exit angle results in a small local slip factor. Reduction of through-flow, which is not of course desirable, increases the slip factor. Larger values of slip factor occur in the reversing flow and stall regions behind the inlet. This has nothing to say about the physics of relative vortical motions or fluid–blade interaction, but it only occurs due to the reduced flow through the rotor.

In a set of experimental data for squirrel-cage fans, deviation was correlated with the space-to-chord ratio and incidence. The extracted correlation allows for a very small share, in the range of 20–50%, for incidence. This is not adequate for squirrel-cage fans, since additional parameters such as the inlet geometry and the volute spread angle control the volumetric flow rate and therefore the incidence.

The slip factor for two fans with different volute spread angles ($\alpha_s = 3°$ and $\alpha_s = 7°$) are obtained from velocity measurements at section $\theta = 360°$ (Figure 4.7). A combination of small volute spread angle ($\alpha_s = 3°$) and small blade exit angle (fan 1 in Figure 4.7) has resulted in slip factors close to unity across the fan. Large volute spread angle ($\alpha_s = 7°$) and large blade exit angle (fan 2 in Figure 4.7) have reduced the slip factor to less than one-fifth almost across the entire rotor width. Larger values behind the inlet just mean no flow and no blade congruency.

Figure 4.8 shows variations in the slip factor at a single cross-section ($\theta = 270°$) for different operating conditions. The measurements are on the same fan with 3° volute spread angle and with outward inlet. This figure shows that reduction of fan flow rate decreases the slip factor.

Figure 4.9 compares axial variation of slip factor for four different circumferential locations for a fan with 7° spread angle. This fan has small slip factors round the rotor.

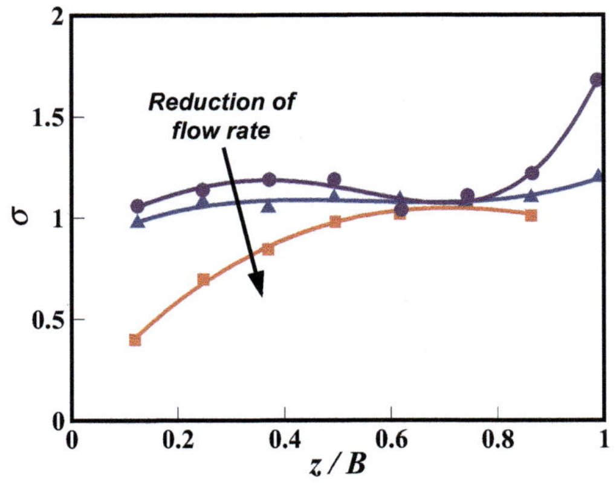

Figure 4.8 Variation of slip factor with fan flow rate.

The values increase behind the inlet, but this is just because there is little, if any, flow through the rotor in that region. Otherwise, the slip factor has no considerable functionality from the circumferential position.

Blade-to-blade variation of velocity components plus slip factor for the same fan ($\alpha_s = 7°$) is shown in Figure 4.10. The cut-off changes jet and wake patterns between the blades. At the two stations just after the cut-off (Figure 4.10(a) and (b)) jets have smaller slip factors, and therefore do not follow the blades. Although there is fluid flow through the rotor after the cut-off, it takes little turning from the blades. Similarly, wakes have larger slip factors but just because there is little through-flow. In the station adequately away from the cut-off (Figure 4.10(c)), jet and wake patterns are more distinguishable and the slip factor remains uniform. This characteristic persists until the cut-off.

The overall impression is that slip factor widely varies between different squirrel-cage fans, and for a single fan at different locations and operating points. It does not

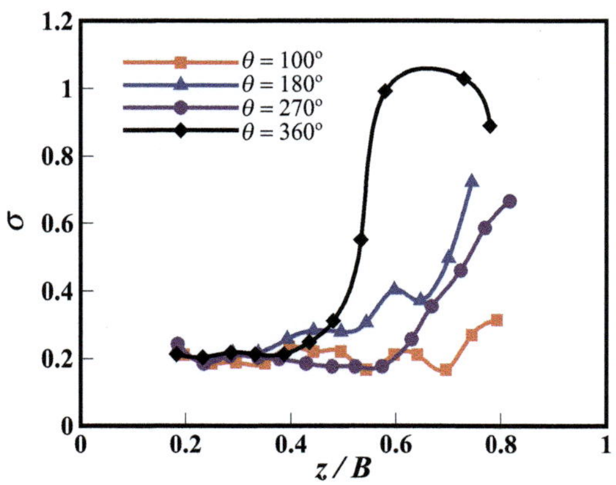

Figure 4.9 Axial variation of slip factor for a fan with $\alpha_s = 7°$. Measurements are at maximum efficiency point ($\Phi = 0.632$) and four different circumferential locations.

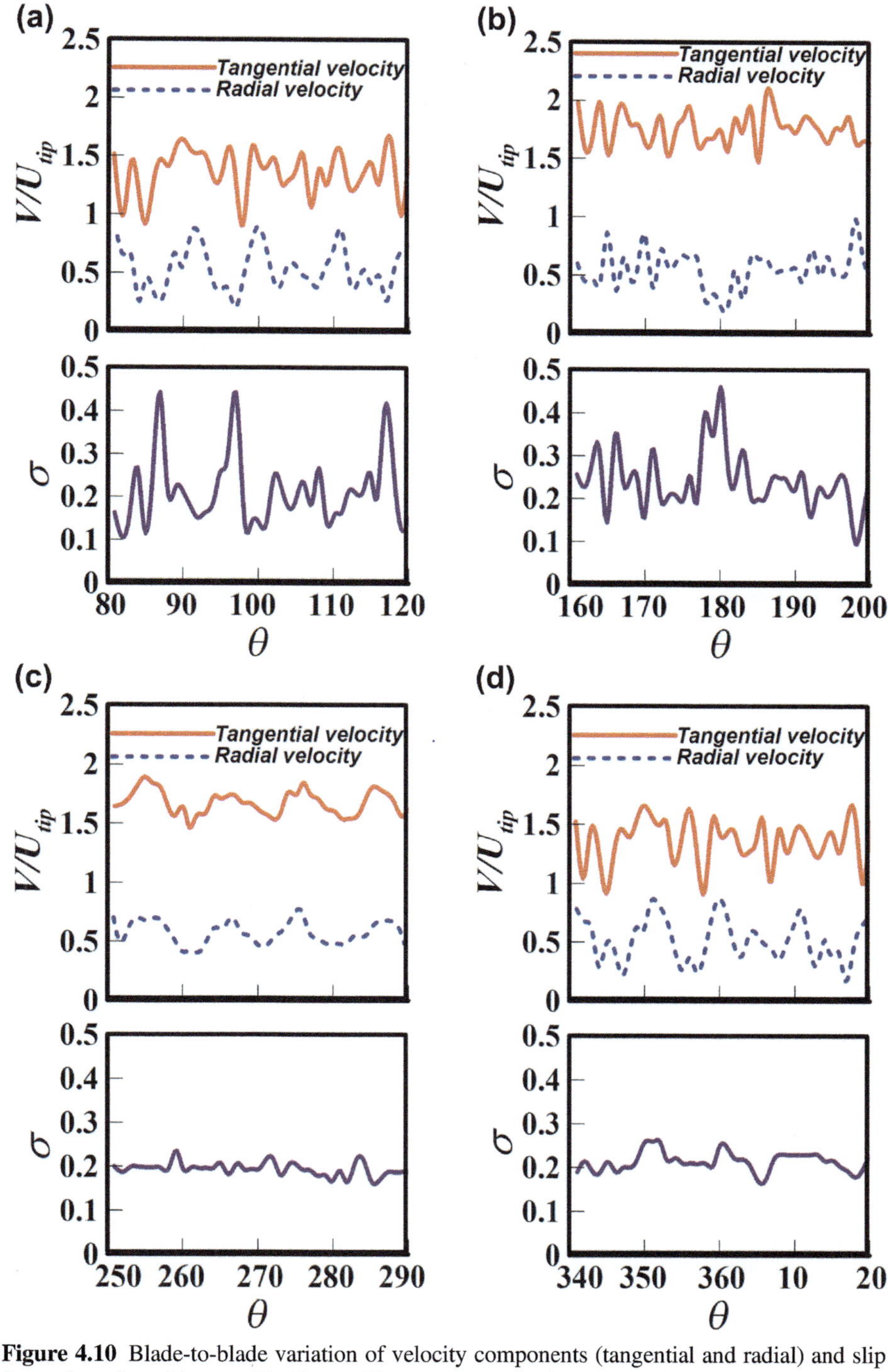

Figure 4.10 Blade-to-blade variation of velocity components (tangential and radial) and slip factor for a fan with $\alpha_s = 7°$. The profiles are shown for four circumferential ranges: (a) $80 \leq \theta \leq 120$; (b) $160 \leq \theta \leq 200$; (c) $250 \leq \theta \leq 290$; (d) $340 \leq \theta \leq 360$ and $0 \leq \theta \leq 20$.

seem to be a complete drawback if the flow does not adhere to the blades in such fans. As the slip factor is coupled with the radial velocity, large deviations from blade exit angle can be due to an increase of flow rate which is beneficial in this case.

4.3 Volute optimization

4.3.1 Optimum volute shape

Until now only variations of individual parameters have led to experimental optimization of the squirrel-cage fan. The approach for squirrel-cage fan sizing starts from the impeller diameter and then other dimensions are defined, normalized and optimized accordingly. Unlike other centrifugal machines, different performances are not obtained through a modification in impeller relative dimensions or blade angles, but are accessible with a change of impeller overall size and speed as well as volute profile and spread angle.

The theoretical method for design of a spiral volute is based on conservation of mass and momentum. Calculation of flow direction in a simple volute assumes a vortex as the flow production source. The effect of the vortex is replaced by an impeller of diameter D, and each streamline can be a volute boundary. Streamline angle (volute spread angle) remains constant along the volute in this approach. The two proposed equations for the volute radius (r) are as follows:

$$r = r_0 \exp(\theta \tan \alpha_s) \tag{4.1}$$

$$r = r_{360^\circ} + (r_{65^\circ} - r_{360^\circ})\exp(-m\theta) \tag{4.2}$$

where θ is the circumferential coordinate, r_{360° and r_{65° are the volute radius at $\theta = 360^\circ$ and $\theta = 65^\circ$, respectively. The volute spread angle is not directly present in the latter equation and the parameter m is used instead of α_s. Profile 1 in Figure 4.11 is based on Eqn (4.1) for $\alpha_s = 5^\circ$ while Profiles 2 and 3 are based on Eqn (4.2) with $m = 0.5$ and $m = 1$, respectively. Parameters such as beginning angle of the volute and exit throttle size are considered to be similar, while the volute boundary curves are different (flow direction in these volutes is expected to be counter-clockwise).

Figure 4.12 illustrates characteristic curves of three fans with the above volutes. The fan that is constructed with Profile 1 shows higher performance and efficiency as compared to the other two, especially at larger flow rates. Additionally, a comparison between fans with Profiles 2 and 3 shows that larger m in Eqn (4.2) results in a fan with poor characteristics.

Volutes with variable cross-section (conical volute) can be used in order to provide more space for the zone with higher flow rate (the major portion of fluid flow occurs from the two-third width of the rotor away from the inlet). A schematic view of this type of volute is shown in Figure 4.13.

Figure 4.14 compares head of fans with one straight and two conical volutes (with 6° of conical angle). Profile 3 (as described above) is considered for the straight volute.

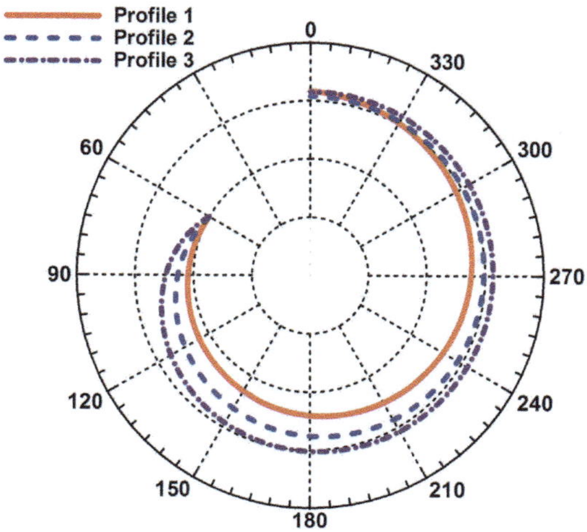

Figure 4.11 Volute profiles with three different boundary curves: Profile 1 based on Eqn (4.1) and Profiles 2 and 3 based on Eqn (4.2).

It is also used for conical volutes; one at the volute back wall (case I) and the other at the middle cross-section of the volute (case II). The conical volute of case I, which has minimum cross-section as compared to other fans, has higher performance than the others.

4.3.2 Performance and aerodynamic flow

Any geometry modification in squirrel-cage fans changes flow field and turbulence characteristics which in turn results in superior/inferior performance or noise characteristic. The research approach that develops from such justification is to study the flow, relate it to geometric parameters and then make relevant conclusions for noise

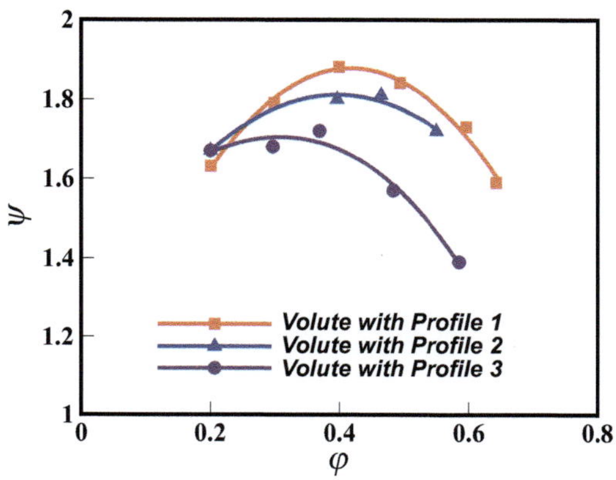

Figure 4.12 Effect of volute profile on head of the fan.

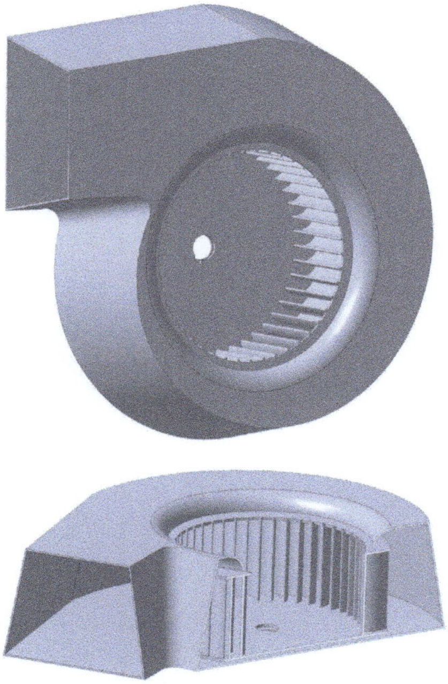

Figure 4.13 Squirrel-cage fan with conical volute.

and performance. This approach can also comply with the Euler energy equation, in which blade angles control the developed head and performance.

The literature on squirrel-cage fan research does not link the above sequence. Although there is an advantage that for a constant impeller diameter, a larger volute spread angle produces more flow, flow field studies do not confirm better aerodynamic characteristics for such configuration. The first evidence for this statement is from measurements of slip factor for squirrel-cage fans that result in larger slip factors for fans

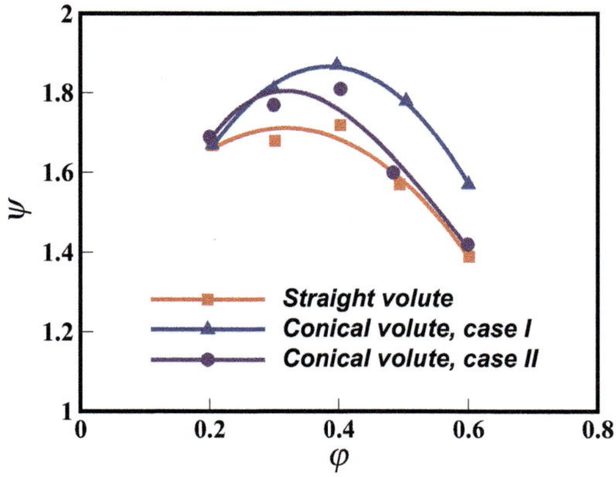

Figure 4.14 Effect of conical volutes on the head coefficient of the fan.

with smaller volute spiral angles. This shows the interdependence of volute and the rotor. It means that for a large value of volute spread angle that is superior in producing larger flow rates, the short blade chord of squirrel-cage fan is less able to guide the flow.

The second evidence is that the performance of a squirrel-cage fan with a larger volute spread angle (and therefore a larger flow rate) is less sensitive to modifications of the inlet shape or the impeller inlet spacing. It was shown in Chapter 3 that the outward-curved inlet nozzle results in a uniform flow at the impeller exit with no inlet flow separation. This improvement does not occur for the volutes with spiral angle larger than 8°. The optimization of other inlet geometries such as the inlet gap, nozzle diameter and its curvature is also effective for smaller volute spread angles and does not occur otherwise.

The conclusion is that a selection of the impeller that is followed by an appropriately sized volute might not result in an aerodynamic flow and therefore is inefficient. Since there is a strong interaction between the inlet, the impeller and the volute, only a joint sizing of these three parts would result in optimization of the fan. A condition that opens this point of view is the selection of a fan for an application with overall space limitation. When the total size is defined, two alternatives are possible:

1. a fan with a smaller volute spread angle and a larger impeller;
2. a fan with a larger volute spread angle and therefore a smaller impeller.

The required flow rate could therefore be developed through a larger rotor tangential velocity combined with a small volute spread angle or through a larger volute with a small impeller radius. The selection could only be resolved through comparative rotor/volute sizing of the fan. This is done through the introduction of a proper length scale that highlights volute as well as different impeller dimensions. The correct combination of rotor and volute length scales can lead to a greater aerodynamic flow and less noise, and can produce a more efficient fan, with no loss of the volumetric flow rate.

4.3.3 Casing-based nondimensional coefficients

It is common to employ rotor-based nondimensional coefficients to compare different designs of a fan type. The flow rate and head coefficients for squirrel-cage fans are then defined as follows:

$$\varphi = \frac{Q}{\pi D_2 b(\pi n D_2)} \tag{4.3}$$

$$\psi = \frac{\Delta P}{\frac{1}{2}\rho(\pi n D_2)^2} \tag{4.4}$$

The impeller diameter appears as a length scale in the reference velocity and the reference flow area. There is no doubt that this diameter is a representative length scale, especially for axial flow-type machines, as it defines the effective size of the inlet, the impeller and the outlet channels.

It has been argued previously that in squirrel-cage fans the volute geometry and the impeller diameter have comparable effects. Nevertheless, when squirrel-cage fan performance is compared through the above nondimensional flow and head coefficients, the volute size is treated as a secondary parameter. A proposal is to use a characteristic diameter based on the two perpendicular casing diameters (Figure 4.1) as a new length scale:

$$D_c = \sqrt{D_{c1}D_{c2}} \tag{4.5}$$

This characteristic diameter represents the contribution of the impeller diameter and the average height of volute channel. Casing-based nondimensional coefficients are therefore defined instead of the impeller diameter and used to calculate flow rate and head coefficients:

$$\varphi_c = \frac{Q}{\pi D_c b(\pi n D_c)} \tag{4.6}$$

$$\psi_c = \frac{\Delta P}{\frac{1}{2}\rho(\pi n D_c)^2} \tag{4.7}$$

The denominators of these quantities are proportional to overall fan size, and therefore each of these nondimensional coefficients can be interpreted as the volume effectiveness of the fan. This is a meaningful concept as it compares fan performance curves on an overall size basis. Large φ_c means more flow is displaced, and large ψ_c means higher head is developed by the fan in a given space.

4.3.4 Application and discussion

It is possible to further develop these definitions when performance curves of different fans are compared. Figure 4.15 shows characteristic curves for a number of different fan configurations that are drawn using the classic definition of nondimensional head and flow rates (Eqns (4.3) and (4.4)). These fans have similar inlet nozzles and impellers while the volute dimensions are different.

Figure 4.16 shows the same performance results that are redrawn using the casing-based normalized coefficients (Eqns (4.6) and (4.7)). The range of variation of flow coefficient in this figure is almost the same, irrespective of the volute spiral angles. It means that total fan size is the controlling parameter for flow rate, and therefore in order to correlate the performance data, the casing-based coefficient of Eqn (4.6) is superior to the impeller-based (classic) coefficient of Eqn (4.3).

Figure 4.17 shows a reproduction of a published data for a simple rotor with volutes of different spread angles employing coefficients of Eqns (4.3) and (4.4). It might be concluded that a wider spread angle has an advantage since it produces a larger flow

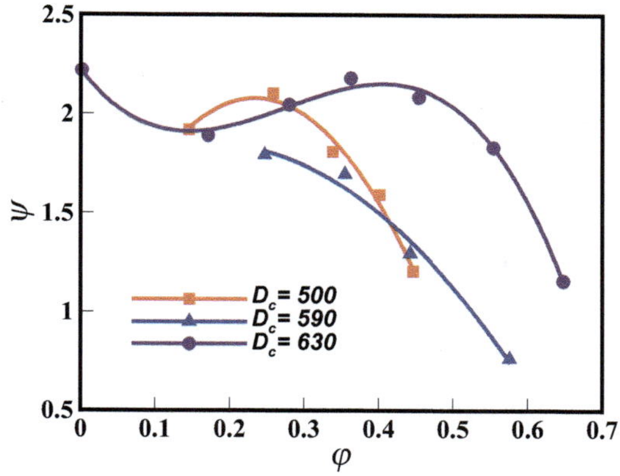

Figure 4.15 The fan performance curves for three volute size according to the classic definitions of head and flow rate coefficients (Eqns (4.3) and (4.4)).

rate,' but this overlooks the drawback of a larger size and a less aerodynamic flow. Figure 4.18 is the reproduction of the same data using the coefficients of Eqns (4.6) and (4.7). Two points should be noted here:

1. It is evident from Figure 4.18 that for all volute spread angles, flow coefficients that are based on casing size fall in a band of up to 0.2. The operational range is within 0.08–0.16 and maximum efficiencies are expected to occur at a flow coefficient of about 0.1.
2. A volute spread angle of 5° offers the largest head and efficiency in the available space. It is important to note that this optimum is achieved with no loss in flow rate since for the same overall space, a larger rotor with the same rotational speed would compensate the reduced flow rate due to a smaller volute spread angle. In this case, the slip factor out of the impeller would be larger and the performance is more susceptible to geometric optimizations.

As mentioned before, the casing-based normalized coefficients utilized both of the volute and rotor dimensions. They are therefore more potent in determining the characteristics of a squirrel-cage fan. For instance, Figure 4.19 illustrates the total head

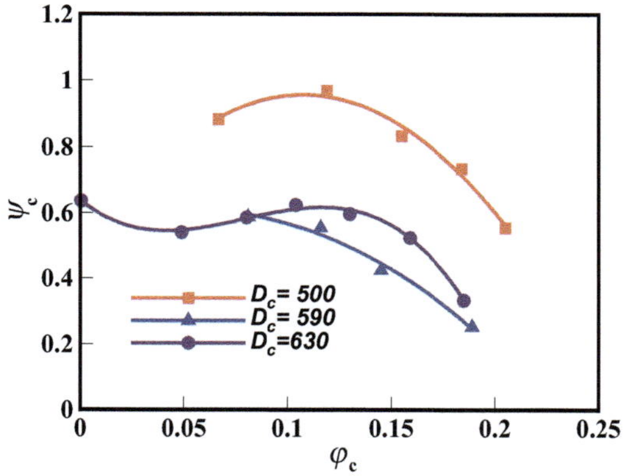

Figure 4.16 The fan performance curves for three volute size according to the casing-based normalized coefficients (Eqns (4.6) and (4.7)).

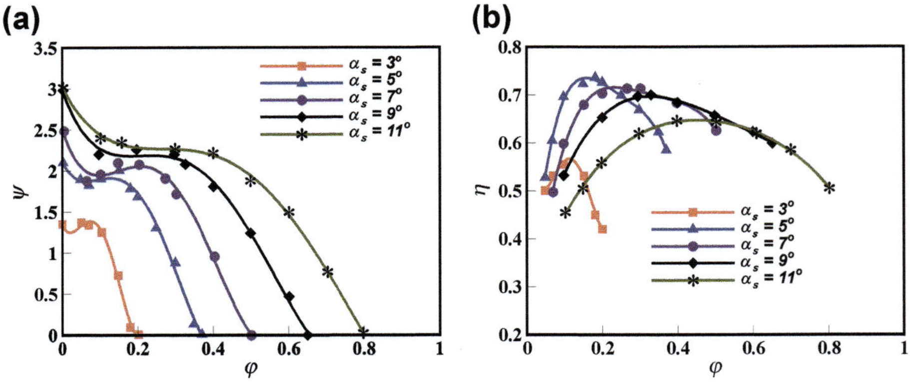

Figure 4.17 Fan performance curves for different volute spread angles using classic definition of the flow coefficient: (a) static head coefficient after Roth (1981) and (b) efficiency.

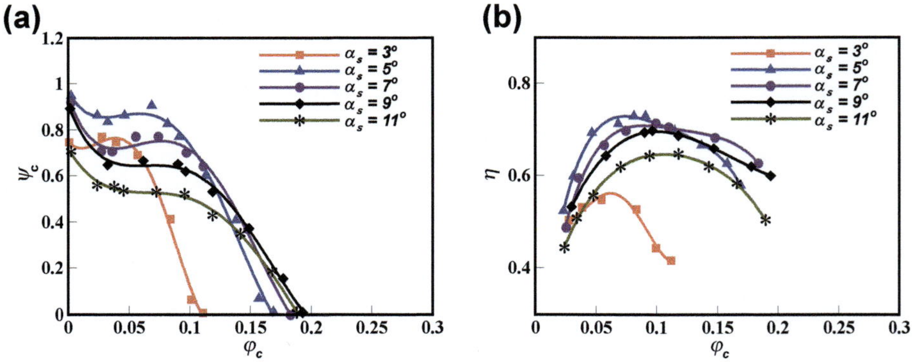

Figure 4.18 A comparison between squirrel-cage fan performance curves using casing-based normalized coefficients: (a) static head and (b) efficiency.

curves of five fans with different volute spread angles. According to Figure 4.19(a), which uses the classic flow and head coefficients, no obvious difference can be observed in curves for $\alpha_s \geq 5°$. On the other hand, utilization of casing-based coefficients in Figure 4.19(b) provides a clear comparison of the characteristic curves in order to determine the optimal fan.

4.3.5 Justification for an optimum volute-to-impeller diameter ratio

The argument for a joint rotor/volute aerodynamic dimensioning of the squirrel-cage fan led to optimum characteristics in the previous section. A comparison of total head coefficients in Figure 4.19 shows that the optimum value for volute spread angle is about 5° that corresponds to a volute-to-impeller length scale ratio of $D_c/D_r = 0.69$.

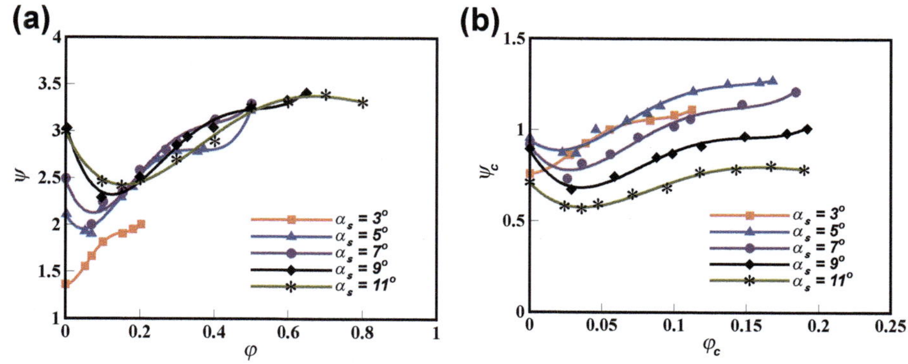

Figure 4.19 The advantage of the new definition of nondimensional head in determining an optimum for the volute spread angle: (a) rotor-based (classic) head coefficient and (b) casing-based head coefficient.

This is the ratio for which the space effectiveness of energy exchange inside the fan is the highest. Such configuration results in a volute-to-impeller width ratio of 1.08.

It is known hitherto that a larger volute spread angle is advantageous as far as it produces a larger flow coefficient and an actually higher flow rate but for a bigger allocated space. For a forward-curved squirrel-cage fan, the impeller occupies a narrow annular space. The volute spiral angle changes the relative volume of the spiral zone after the impeller to that of the cylindrical inlet zone. When a smaller volute spiral angle is considered in a given space for the fan, the larger developed head coefficient is primarily due to a larger relative available space of the impeller. A larger slip factor is also obtained when this relatively smaller spiral angle is employed.

For a smaller volute spiral angle than the optimum, the outlet space after the impeller is relatively small. The dominant friction resistance in such a volute blocks the flow, and the recirculating flow back into the rotor reduces head and efficiency.

For a given fan space, if the spiral angle is larger than the optimum, the inlet zone would be relatively smaller. This does not reduce the flow into the fan since the performance results show that the casing-based flow coefficient has a constant range and the flow rate only depends on the overall size. Higher velocity and vacuum at the entrance of such a fan result in a wider inlet separating zone and more nonuniform velocity profile after the blades. Again such dissipative phenomenon leads to energy loss in the volute and reduces fan head and efficiency.

4.3.6 Application to design

The advantage that 'the casing-based normalized flow rate has the same range for all selections of volute spread angle' leads to another positive point. The maximum head coefficient of Eqn (4.6) now occurs at almost similar flow coefficient for different squirrel-cage fans. A similar argument is true for maximum efficiency. When the available space is fixed through the total casing size, the impeller speed can be found from Eqn (4.6), as the required flow rate is known. The required head and efficiency would then be used to determine the impeller diameter and the volute spread angle. The following procedure is therefore envisaged for selecting the impeller diameter when the available space is a constraint:

1. Define and evaluate the available space.
2. Use the optimum value for the casing-based normalized flow coefficient to select the speed and fulfil the required flow.
3. Select the optimum volute spread angle.
4. Find the impeller diameter for the optimum cut-off clearance and impeller diameter ratio.
5. Extract the head from the curves of the implemented rotor and compare it with the required head.
6. If the required head is lower, a larger spiral angle and smaller impeller diameter can be selected to match the fan with the system.

4.4 Volute width

A known drawback that contributes to low efficiency of the squirrel-cage fan is that flow must turn 90° from the inlet to the rotor. As mentioned in this and previous chapters, there is a separation zone behind the inlet, which occupies about one-third of the rotor, and therefore the bulk of the flow turns near the backplate.

Volute dimensions such as width and spread angle have controlling effects on flow through the rotor. There is disagreement on the selection of relative width of rotor and volute. Some suggest a volute width as close as possible to the rotor width just to allow for inlet lip intrusion and manufacturing tolerance and expect efficiency reduction if B/b (volute-to-rotor width ratio) exceeds 1.2. Others suggest the maximum efficiencies for $B/b = 2.5$ for narrow rotors ($4b/D_1 = 1.0$). Similar to the previous section, there must be a new definition for flow characteristics to account for volute or rotor width.

The general form of the flow coefficient is defined as a function of flow rate, the reference velocity (U_{ref}) and the reference area (A_{ref}):

$$\varphi = \frac{Q}{U_{ref}\,A_{ref}} \tag{4.8}$$

The blade tip velocity is considered as the reference velocity:

$$U_{ref} = \pi DN \tag{4.9}$$

If axial width is intended to appear in fan parameters, flow coefficient can be considered from two different viewpoints, namely, radial and axial flow coefficients. This comes about from two different definitions for the reference area. When the inlet surface is the reference area ($A_{ref} = \pi D^2/4$), the radial flow coefficient is

$$\varphi_r = \frac{4Q}{\pi^2 ND^3} \tag{4.10}$$

This is the commonly used radial flow coefficient that is not characterized by the fan width. If the rotor outlet surface is considered as the reference area ($A_{ref} = \pi bD$), the axial flow coefficient is obtained:

$$\varphi_a = \frac{Q}{\pi^2 D^2 bN} \tag{4.11}$$

The performance curves of four fans with different rotor width-to-diameter ratios are presented in Figure 4.20 versus radial flow coefficient. Fan characteristics in Figure 4.20(a) are classified into two different groups. The first group consists of two fans with narrower volute such that their head coefficients decline fast with flow rate. The second group is relevant to the two remaining fans with wider volute and in this case, larger maximum flow rate and also a more flat characteristic can be observed. Figure 4.20(a) shows that the wider volute allows more flow through the rotor and also flattens the characteristic curve. A change of rotor width within the same volute only marginally changes the maximum flow rate, but the slope of the characteristics remains the same. Efficiency trends in Figure 4.20(b) show that the two fans with the wider rotor (and therefore wider volute) are generally more efficient. The conclusion from these graphs is that the volute width has the prime importance.

There are other experimental results that show how fan performance varies with the ratio between rotor width and rotor diameter when the volute-to-rotor width ratio is constant. Figure 4.21 shows performance curves of fans with different widths as drawn with radial or axial flow coefficients. It shows that wider fans do not pump a correspondingly larger flow. Performance curves for wider rotors are more flat but end at respectively smaller flow rates.

Maximum total pressure coefficient versus rotor width-to-diameter ratio is plotted in Figure 4.22(a). The fan with the ratio of $4b/D_1 = 1.72$ has the largest total pressure coefficient in this range. Fan axial flow rate for maximum head is extracted from Figure 4.21(b) and presented in Figure 4.22(b) versus the ratio of $4b/D_1$.

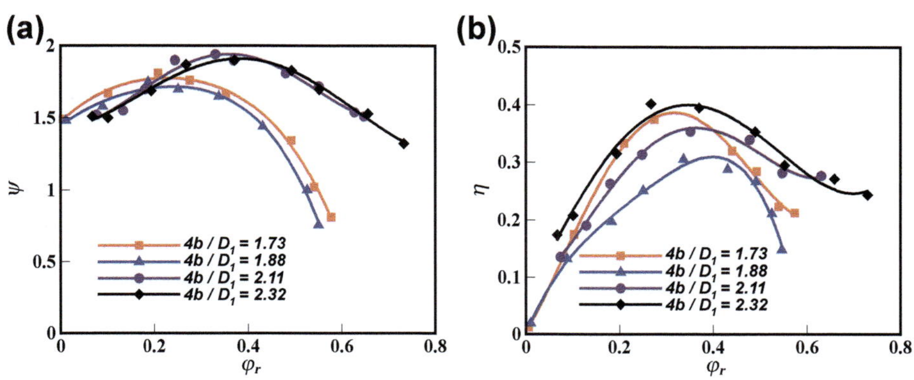

Figure 4.20 Variation of head coefficient (a) and efficiency (b) with radial flow coefficient for different rotor width-to-diameter ratios.

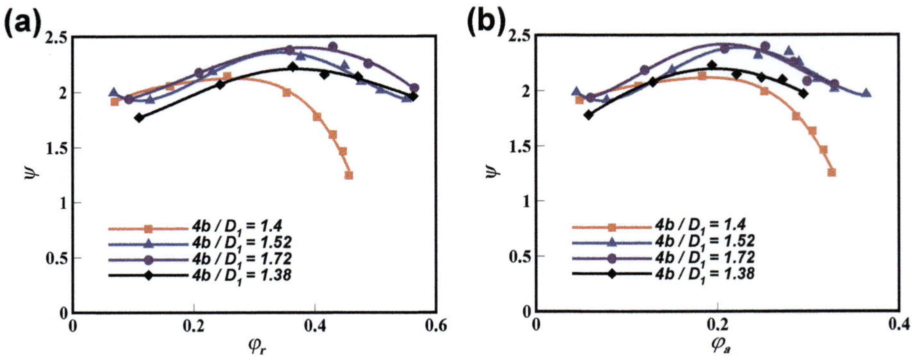

Figure 4.21 Fan head coefficient for various rotor width-to-diameter ratios versus (a) radial flow coefficient and (b) axial flow coefficient.

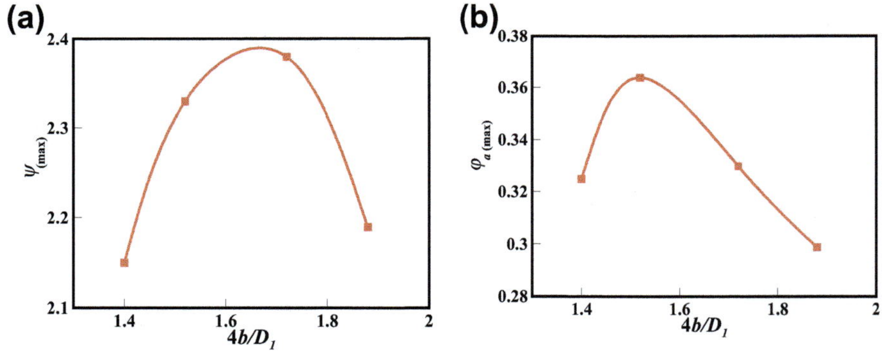

Figure 4.22 Peak performance parameters as a function of rotor width-to-diameter ratio: (a) maximum total pressure coefficient and (b) maximum axial flow rate coefficient.

Larger flow coefficient of fan with $4b/D_1 = 1.52$ shows that this fan can produce high flow rates for a similarly higher occupied volume. The width-to-diameter ratio is an indication of the ratio of rotor outlet to inlet areas. When a similar flow passes through volutes with different width, slower average velocity is expected for a larger width. This does not necessarily mean less energy transfer (as in Figure 4.21(a) the maximum flow rate and head are equal for all fans with the ratio larger than 1.52).

Fans normally operate in the right-hand part of their performance curve. In Figure 4.23, total pressure coefficient for fans with different width-to-diameter ratios at a selected flow coefficient (larger than that of the highest pressure rise) is plotted. The fan with $4b/D_1 = 1.72$ has the highest total pressure coefficient for the same flow rate, but for equal axial flow coefficient (equal occupied volume) the fan with $4b/D_1 = 1.52$ has the highest total pressure.

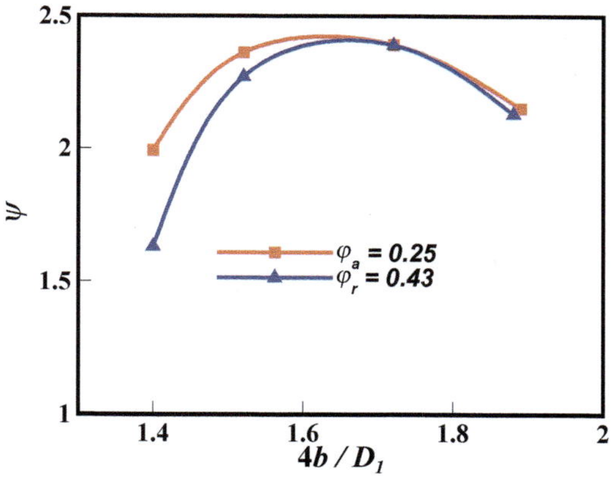

Figure 4.23 Total pressure coefficient versus $4b/D_1$ for constant axial flow coefficient ($\varphi_a = 0.25$) or constant radial flow coefficient ($\varphi_r = 0.43$).

4.5 Double-outlet volute

Centrifugal fans in heating and ventilation industries usually supply more than one outlet port. A central unit in such industries supplies conditioned air to various enclosures through different ports. Forward-curved squirrel-cage fans are the prime candidates for such applications. Automobile ventilation is an example where such fans serve both right and left sides of the cabin.

A double-outlet centrifugal fan has a volute with two outlets. The fan can be installed between the delivery ports where there would be less turning in the channelling after the fan. It is also more flexible regarding its positioning and the required space. A double-outlet fan has two cut-offs. This can be a disadvantage since there is reversing flow and blockage at part-load condition round the cut-off. It is important to minimize this reversing flow.

A double-outlet squirrel-cage fan is shown in Figure 4.24. A double-inlet configuration is also helpful in heating and ventilation industries due to its higher volumetric flow rates. This can be a prime candidate for double-outlet fans.

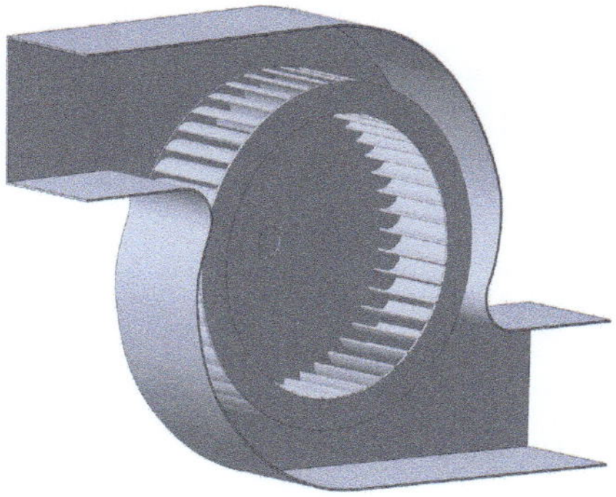

Figure 4.24 A double-outlet squirrel-cage fan.

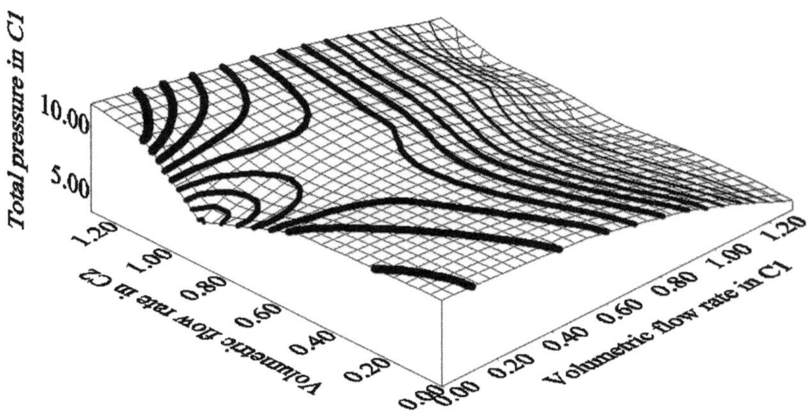

Figure 4.25 Total pressure variation in channel 1 versus volumetric flow rate in both channels (C1 and C2 denotes channels 1 and 2, respectively). The volumetric flow rate and total pressure are measured in m^3/s and millimetre of water, respectively.

The two-outlet channels are termed channels 1 and 2. There are no conclusive differences in the measured data for either channel. The designations 1 and 2 only show which channel is throttled and where the pressure measurements are made.

Figure 4.25 depicts variation of total pressure in channel 1 versus flow rate at each channel, and Figure 4.26 shows the same for channel 2. Variations in the total pressure of both channels are fairly similar. It does not make any difference which of the two outlet channels has a constant throttle point when the other is throttled. Pressure in the fixed throttled channel remains constant when the other channel is throttled. This is important since it shows that if application setting in one channel changes, the users of the other channel will not experience higher or lower heads.

Figure 4.27 shows efficiency of the double-outlet fan versus flow rate at each channel. This figure indicates that non-symmetric use of a double-outlet fan can reduce

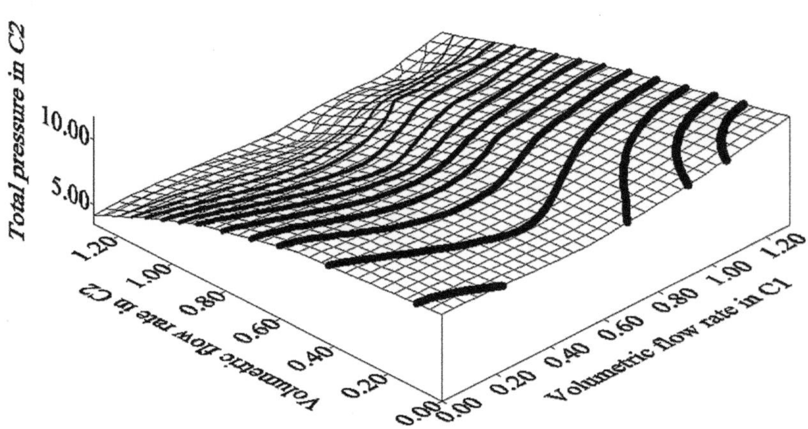

Figure 4.26 Total pressure variation in channel 2 versus flow rate in both channels. The volumetric flow rate and total pressure are measured in m^3/s and millimetre of water, respectively.

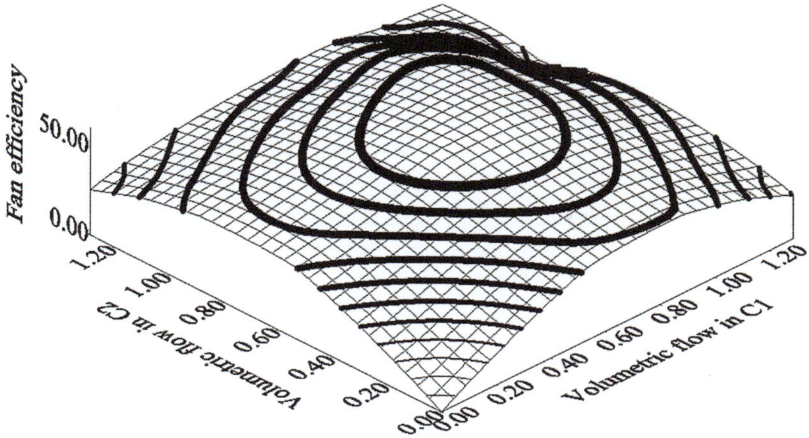

Figure 4.27 Variation of double-outlet fan efficiency versus flow rate in both channels. The volumetric flow rate is measured in m^3/s.

efficiency. In other words, the best efficiency conditions are for symmetric operation. There are some sharp drops in performance, when either of channels is closed. This is not a drawback for this fan. It makes more sense to use a single-outlet fan if only one exit channel is required.

Figure 4.28 shows a comparison between a double-outlet fan and a double-inlet single-outlet fan of similar construction. Three curves show the performance of the double-outlet fan at three different conditions. In these conditions, one channel is set to 0%, 50% and 100% of the maximum flow rate, and the other is open. The figure shows that a double exit fan can produce a higher volumetric flow rate than a single-outlet one when both channels are fully utilized.

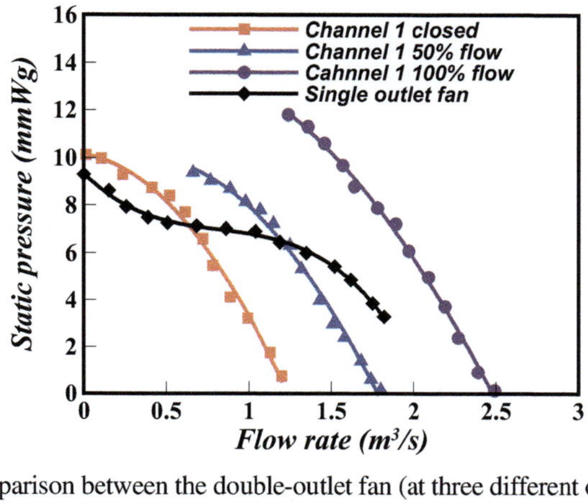

Figure 4.28 A comparison between the double-outlet fan (at three different operating conditions) and a similar industrial single-outlet fan. The gage static pressures are measured in millimetre of water.

4.6 Final remarks

Two important parameters must be considered in selection or design of the volute of a squirrel-cage fan: the volute spreading angle and the volute width. The optimum value of spread angle was found to be 5°. The volute width can be selected according to the results in this chapter. The volute profile is a secondary parameter in squirrel-cage fans and has a minor effect as compared to the first two.

Further reading

Akbari, G., Montazerin, N., & Akbarizadeh, M. (2011). Stereoscopic particle image velocimetry of the flow field in the rotor exit region of a forward-blade centrifugal turbomachine. *Proceedings of the Institution of Mechanical Engineers, Part A: Journal of Power Energy, 226*, 163–181.

Cho, G. R., Kawahashi, M., Hirahara, H., & Kitadume, M. (2005). Application of stereoscopic particle image velocimetry to experimental analysis of flow through multi-blade fan. *JSME International Journal, 48*(1), 25–33.

Damangir, A. (2004). *Joint impeller/scroll sizing of squirrel cage fans using alternative non-dimensional head and flow rate coefficients* (in Persian) (Ph.D. thesis). Amirkabir University of Technology.

Damangir, A., & Montazerin, N. (1998). Stall cells in the volute of a squirrel cage fan. In *Proceedings of the integrated design and process technology conference (IDPT), Berlin, Germany* (pp. 261–267).

Denger, G. R., & McBride, M. W. (1990). Three-dimensional flow field characteristics measured in a forward-curved centrifugal blower using particle tracing velocimetry. In *Proceedings of fluid measurement and instrumentation forum, ASME publication FED, 95* (pp. 49–56).

Dong, R., Chu, S., & Katz, J. (1992). Quantitative visualization of the flow within the volute of a centrifugal pump. Part B: results and analysis. *Journal of Fluids Engineering, 114*, 396–403.

Heshmat Dehkordi, A. (2005). *A study of rotor width effect on squirrel cage fan performance* (in Persian) (MSc. thesis). Amirkabir University of Technology.

Meakhail, T., & Park, S. O. (2005). A study of impeller-diffuser-volute interaction in a centrifugal fan. *Journal of Turbomachinery, 127*, 84–90.

Montazerin, N. (2000). A double-outlet squirrel-cage fan. In *Proceedings of the 5th biennial conference on engineering system design and analysis, ASME.*

Montazerin, N., & Damangir, A. (1996). Flow field measurements in the volute section of a squirrel cage fan. In *Proceedings of the second biennial engineering systems design and analysis conference, Montpellier, France* (pp. 237–257).

Montazerin, N., Damangir, A., & Kazemi Fard, A. (1998). A study of slip factor and velocity components at the rotor exit of forward-curved squirrel cage fans, using laser Doppler anemometry. *Proceedings of the Institution of Mechanical Engineers, Part A: Journal of Power Energy, 215*, 453–463.

Montazerin, N., Damangir, A., & Mirian, S. (1998). A new concept for squirrel-cage fan inlet. *Proceedings of the Institution of Mechanical Engineers, Part A: Journal of Power Energy, 212*, 343–349.

Pedersen, N., Larsen, P. S., & Jacobsen, C. B. (2003). Flow in a centrifugal pump impeller at design and off-design conditions—part I: particle image velocimetry (PIV) and laser Doppler velocimetry (LDV) measurements. *Journal of Fluids Engineering, 125*, 61–72.

Raj, D., & Swim, W. B. (1981). Measurements of the mean flow velocity and velocity fluctuations at the exit of an FC centrifugal fan rotor. *Journal of Engineering for Power, 103*, 393–399.

Roth, H. W. (1981). Optimierung von trommelläufer-ventilatoren. *Strömungsmechanik und Strömungs-maschinen, 29*, 1–45.

Samarbakhsh, S. (2007). *Study on squirrel cage fan volute profile* (in Persian) (MSc. thesis). Amirkabir University of Technology.

Samian, R.S. (2009). *A numerical study of fan width effect on squirrel cage fan performance* (in Persian) (MSc. thesis). Amirkabir University of Technology.

Sinha, M., & Katz, J. (2000). Quantitative visualization of the flow in a centrifugal pump with diffuser vanes—I: on flow structures and turbulence. *Journal of Fluids Engineering, 122*, 97–107.

Sinha, M., Katz, J., & Meneveau, C. (2000). Quantitative visualization of the flow in a centrifugal pump with diffuser vanes—II: addressing passage-averaged and large-eddy simulation modeling issues in turbomachinery flows. *Journal of Fluids Engineering, 122*, 108–116.

Noise in forward-curved centrifugal fans

5

5.1 Sound parameters

The concepts of sound power, energy and pressure are essential in understanding the sound radiation and emission. This section introduces some of the basic acoustical parameters that are important in the study of fan noise and implementation of experiments in this field.

5.1.1 Sound pressure

Sound is associated with local fluctuation of medium particles, which is composed of local compression rapidly followed by a local rarefaction. Sound travels in longitudinal waves where the small local particle displacement is in the same direction as the wave movement. Therefore, sound travels as waves of compressions and rarefactions with associated pressure variations. The acoustic pressure is the difference between the pressure in the presence of the sound and in the absence of any sound at any point. The acoustic overpressure varies sinusoidally, and it is convenient to use its root mean square (*rms*) value, which is defined as the square root of the average of the squares of the instantaneous pressures:

$$p_{rms} = \left(\frac{1}{T} \int_0^T p^2 \, dt \right)^{1/2} \tag{5.1}$$

The sound spectrum of fans covers a continuous band of frequencies in practice. The following relation is therefore applicable for the *rms* of total pressure ($p_{t,\ rms}$):

$$p_{t,rms} = \left(\int p_{rms}^2 \, df \right)^{1/2} \tag{5.2}$$

5.1.2 Sound power and sound intensity

Any sound source radiates power in the form of sound transmission. When the sound power W passes through an area A, the sound intensity I is defined as power per unit area:

$$I = \frac{W}{A} \tag{5.3}$$

Developments in Turbomachinery Flow.

The sound intensity can be related to sound pressure:

$$I = \frac{p_{rms}^2}{\rho c} \tag{5.4}$$

where ρ is the medium density and c is the speed of sound in that medium.

5.1.3 Sound level

The values of sound pressure, intensity and power are measured on a logarithmic scale, that is, the decibel scale. The decibel scale of the sound is a ten-based logarithmic ratio that can be defined for the sound-power level (SWL), the intensity level (IL) and the sound-pressure level (SPL):

$$\text{SWL} = 10\log\left(\frac{W_1}{W_0}\right) \tag{5.5}$$

$$\text{IL} = 10\log\left(\frac{I_1}{I_0}\right) \tag{5.6}$$

$$\text{SPL} = 10\log\left(\frac{p_1^2}{p_0^2}\right) = 20\log\left(\frac{p_1}{p_0}\right) \tag{5.7}$$

where the constants $W_0 = 10^{-12}\ w$, $I_0 = 10^{-12}\ \text{wm}^{-12}$ and $p_0 = 20\ \mu\text{Pa}$ are utilized.

5.2 Different types of noise

Fan noise is composed of a large number of sound waves, which are periodic, sinusoidal and have constant amplitudes. In the study of fan noise, the corresponding "frequency spectrum" associated with sound waves is a useful tool for sound generation analysis. The spectrum is an illustration of the frequency content and the corresponding magnitude of a sinusoidal signal. The spectral diagram shows the effect of discrete frequency phenomena and the bandwidth background noise. A sample frequency spectrum for a squirrel-cage fan is shown in Figure 5.1.

Fan noise consists of vibration-induced and flow-induced modes. Flow-induced noise, which is also termed aerodynamic noise, is generated by aerodynamic forces and motions, while the vibration-induced noise is generated by vibrations of fan components. Aerodynamic noise has three origins: (1) quadruple noise, related to turbulence shear stress; (2) dipolar noise, produced by steady and unsteady forces from solid surfaces on the flow and (3) monopole (thickness noise), due to volume displacement of the moving surfaces. Forces of turbulent shear flow acting on the blades, vanes and casing are the main contributors to fan noise. Forces on blades

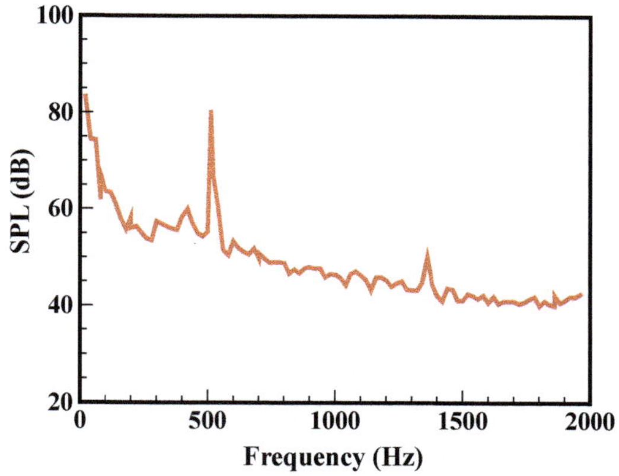

Figure 5.1 A frequency spectrum of the sound of a squirrel-cage fan.

can be periodic or random, and hence the resulting sound field will have either discrete or broadband components. Discrete components due to this mechanism are produced at the blade passing frequency, BPF ($\sim$500 Hz in Figure 5.1) and their harmonics. This is termed tonal noise. Discrete tonal noise is generated due to the interaction between the rotor exit jet and wake flow and the volute cut-off.

The BPF is the most important noise source in centrifugal fans. Broadband components result from inlet turbulence and vortex shedding at blade trailing edges. Another discrete noise in frequency spectrum of the fan is at the rotation frequency (RF). This noise is of vibration-induced type, and in the case of squirrel-cage fans is due to vibration of the volute casing (RF$\approx$10 Hz in Figure 5.1).

5.3 Modelling of sound generation in fans

Sound generation and propagation in centrifugal fans is generally simulated using the hybrid aeroacoustic methods, in which the flow and acoustic filed are separately modelled. The fluid flow is solved first in this method, noise sources are then specified using the simulated flow field, and finally the acoustic field is calculated.

Hybrid aeroacoustic methods are based on Lighthill acoustic analogy that starts from the conservation of mass and momentum:

$$\frac{\partial \rho}{\partial t} + \frac{\partial \rho u_i}{\partial x_i} = 0 \tag{5.8}$$

$$\frac{\partial \rho u_i}{\partial t} + \frac{\partial \rho u_i u_j}{\partial x_j} = -\frac{\partial p}{\partial x_i} + \frac{\partial \tau_{ij}}{\partial x_j} \tag{5.9}$$

where τ_{ij}, ρ, u_i and p are the viscous stress term, density, velocity and pressure of the fluid, respectively. The effect of body force is neglected in this version of the momentum equation. After taking the partial derivative of Eqn (5.8) with respect to time, subtracting it from the divergence of Eqn (5.9) and some manipulation, the following equations are obtained which are nonhomogeneous Lighthill equation for density fluctuations:

$$\frac{\partial^2 \rho'}{\partial t^2} - c_0^2 \frac{\partial^2 \rho'}{\partial x_i^2} = \frac{\partial^2 T_{ij}}{\partial x_i x_j} \tag{5.10}$$

$$T_{ij} = \rho u_i u_j + \delta_{ij}\left(p' - c_0^2 \rho'\right) - \tau_{ij} \tag{5.11}$$

where $\rho' = \rho - \rho_0$ and $p' = p - p_0$ are density and pressure fluctuations, respectively (ρ_0 and p_0 are the density and pressure in the absence of sound waves). T_{ij}, δ_{ij} and c_0 are the Lighthill tension tensor, the Kronecker delta function and the sound speed, respectively.

Viscous and compressibility terms can be neglected, which means that the viscosity and conduction have a minor effect on wave energy dissipation:

$$\frac{\partial^2 T_{ij}}{\partial x_i x_j} \approx \frac{\partial^2 \rho u_i u_j}{\partial x_i x_j} \tag{5.12}$$

Using the free space Green function, the integral form of the Lighthill equation is obtained:

$$\rho'(x,t) = \frac{1}{4\pi c_0^2} \frac{\partial^2}{\partial x_i x_j} \int_{\Omega} \frac{T_{ij}(y, t - |x-y|/c_0)}{|x-y|} \mathrm{d}y \tag{5.13}$$

where Ω is the domain in which noise is generated due to turbulence and x and y are the vectors of noise receiver position and noise source position, respectively. Solving Eqn (5.13) results in fluctuations of sound density at location x and in time t.

The integral equation obtained by the Lighthill equation as the solution of the inhomogeneous equation is in the absence of any solid plate or body in the flow field. Ffowcs Williams and Hawkings develop the equation to predict the sound generated by fluid flow in the presence of arbitrarily moving bodies, which is known as FW–H equation. The first step to add the effect of solid bodies is to develop the Lighthill equation for an unbounded space that is partitioned by mathematical surfaces that exactly correspond to real surfaces.

Figure 5.2 depicts a fixed space of volume Ω enclosed by a surface Γ. The volume Ω is divided into regions 1 and 2 by a discontinuity surface $\Gamma_{1,2}$. Figure 5.2 shows that surface $\Gamma_{1,2}$ is moving in region 2 with velocity v. It should be noted that $\Gamma_{1,2}$ may consist of several closed surfaces. The normal vector of $\Gamma_{1,2}$ from region 1 to region 2 is denoted by n.

The conservation of mass and momentum equations in the presence of the discontinuity surface $\Gamma_{1,2}$ is expressed as follows:

$$\frac{\partial \overline{\rho}}{\partial t} + \frac{\partial \overline{\rho u_i}}{\partial x_i} = \rho_0 v_i \delta(f) \frac{\partial f}{\partial x_i} \tag{5.14}$$

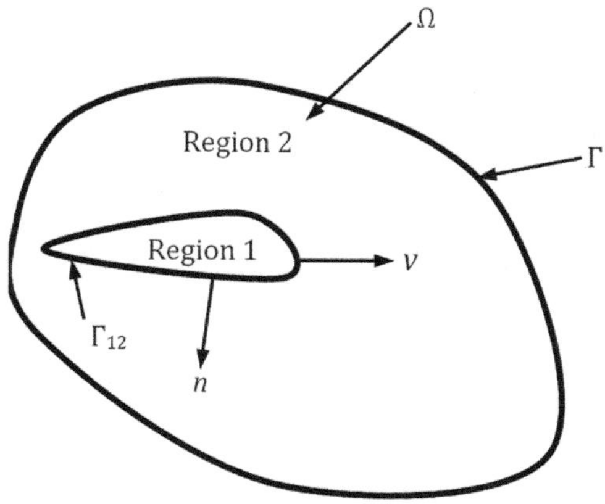

Figure 5.2 An unbounded region with a moving solid body.

$$\frac{\partial \overline{\rho u_i}}{\partial t} + \frac{\partial \left(\overline{\rho u_i u_j} + \overline{p_{ij}}\right)}{\partial x_j} = p_{ij}\delta(f)\frac{\partial f}{\partial x_i} \tag{5.15}$$

where ρ_0 is the fluid density in the absence of sound wave, v_i is the velocity of the discontinuity surface and $\delta(f)$ is the delta function. In Eqns (5.14) and (5.15), $f = 0$ defines the surface $\Gamma_{1,2}$, and negative and positive values of f define regions 1 and 2, respectively. $\overline{\rho}$ is equal to ρ_1 in region 1 and ρ_2 in region 2. Therefore,

$$\overline{\rho} = \begin{cases} \rho & f > 0 \\ \rho_0 & f < 0 \end{cases} \tag{5.16}$$

$$\overline{\rho u_i} = \begin{cases} \rho u_i & f > 0 \\ 0 & f < 0 \end{cases} \tag{5.17}$$

$$\overline{p_{ij}} = \begin{cases} (p - p_0)\delta_{ij} & f > 0 \\ 0 & f < 0 \end{cases} \tag{5.18}$$

By implementing the steps in derivation of Eqn (5.10), the FW–H equation for fluctuations of density is obtained:

$$\frac{\partial^2 \rho'}{\partial t^2} - c_0^2 \frac{\partial^2 \rho'}{\partial x_i^2} = \frac{\partial^2 \overline{T_{ij}}}{\partial x_i x_j} - \frac{\partial}{\partial x_i}\left(p_{ij}\delta(f)\frac{\partial f}{\partial x_i}\right) + \frac{\partial}{\partial t}\left(\rho_0 v_i \delta(f)\frac{\partial f}{\partial x_i}\right) \tag{5.19}$$

$$\overline{T_{ij}} = \begin{cases} \rho u_i u_j + p_{ij} - c_0^2 \rho' \delta_{ij} & f > 0 \\ 0 & f < 0 \end{cases} \tag{5.20}$$

The first term in the right-hand side of Eqn (5.19) is associated with quadruple noise, while the second and third terms are responsible for dipole and monopole, respectively. In most applications of the subsonic fluid flow, especially the fan noise problems, the quadruple term may be neglected. After this assumption, the FW–H equation is simplified as follows:

$$\frac{\partial^2 \rho'}{\partial t^2} - c_0^2 \frac{\partial^2 \rho'}{\partial x_i^2} = \frac{\partial}{\partial x_i}\left(p_{ij}\delta(f)\frac{\partial f}{\partial x_i}\right) + \frac{\partial}{\partial t}\left(\rho_0 v_i \delta(f)\frac{\partial f}{\partial x_i}\right) \tag{5.21}$$

Equation (5.21) is an inhomogeneous wave equation for fluid-flow problems in an unbounded free space. Using the free space Green function, $\delta(g)/4\pi r$, with $g = \tau - t + r/c$ and $r = |x - y|$, the integral form of Eqn (5.21) is obtained as follows:

$$4\pi p'(x,t) = \frac{1}{c}\frac{\partial}{\partial t}\int\limits_{f=0}\left(\frac{\rho_0 c v_n + l_r}{r|1 - M_r|}\right)_{ref} \mathrm{d}S + \int\limits_{f=0}\left(\frac{l_r}{r^2|1 - M_r|}\right)_{ret} \mathrm{d}S \tag{5.22}$$

$$l_i = p_{ij} n_j \tag{5.23}$$

where r is the vector from the observer to the noise source, v_n is the local normal velocity of source surface and M_r is the local Mach number vector of source with respect to a frame that is fixed to the undisturbed medium in r direction. Also $l_r = l_i r_i$ where l_i is the component of local force intensity that acts on the fluid ($l_i = p_{ij} n_j$). The subscript ret is the retarded time ($\tau = t - r/c$) and the integration is along the solid surface ($f = 0$). Equation (5.22) has the advantage that spatial differentiation is avoided; however, the time derivative of the first integral must be performed. The speed and accuracy of noise simulations are improved by eliminating the time derivation. The definition of the retarded time gives

$$\left.\frac{\partial}{\partial t}\right|_X = \frac{1}{1 - M_r}\left.\frac{\partial}{\partial \tau}\right|_x \tag{5.24}$$

The $|_X$ means that the differentiation is performed for fixed position of the observer. Using this definition, the following equations for simulation of flow-induced sound emission are obtained:

$$p'(x,t) = p'_T(x,t) + p'_L(x,t) \tag{5.25}$$

$$p'_T(x,t) = \frac{1}{4\pi}\int\limits_{f=0}\frac{\rho_0(v_n + \dot{v}_n)}{r|1 - M_r|^2}\mathrm{d}\Gamma + \frac{1}{4\pi}\int\limits_{f=0}\frac{\rho_0 v_n\left(r\dot{M}_r + cM_r - cM^2\right)}{r^2|1 - M_r|^3}\mathrm{d}\Gamma \tag{5.26}$$

$$p'_L(x,t) = \frac{1}{4\pi c}\int_{f=0} \frac{\dot{l}_r}{r|1-M_r|^2} d\Gamma + \frac{1}{4\pi}\int_{\Gamma} \frac{l_r - l_M}{r|1-M_r|^2} d\Gamma + \frac{1}{4\pi c}\int_{\Gamma} \frac{l_r\left(r\dot{M}_r + cM_r - cM^2\right)}{r^2|1-M_r|^3} d\Gamma \tag{5.27}$$

where the dot symbol in the above equations represents the time derivation and n is the unique vector normal to the surface. The above system of equations is known as Farassat 1A formulation that is utilized in most numerical works on prediction of sound in centrifugal fans.

It is assumed in simulation of flow-induced noise using the FW−H equation that the sound emission occurs in an unbounded free space. Since the volute scatters the sound wave, the boundary element method (BEM) is used to consider the scattering effect of solid bodies on sound-radiation problems. The conventional BEM requires that the solid body should be a closed or a thick body. Hence employment of standard BEM to predict sound-scattering effect of a volute casing, which is a thin body, results in thin-shape breakdown. The thin-body BEM can overcome this drawback. The computations in this method are in frequency domain. The generated sound by the fan and the received sound at any position in the space are composed of two components, namely, the incidence and the scattered acoustic pressure:

$$p'(x,f) = p'_I(x,f) + p'_S(x,f) \tag{5.28}$$

where $p'_I(x,f)$ and $p'_S(x,f)$ are the incidence and scattered sound pressures. The incidence sound pressure is calculated as follows:

$$p'_I(x,f) = \int_{\Gamma}\int_{\tau} p'[y(\tau),\tau]\frac{\partial G[x,y(\tau),f]}{\partial n[y(\tau)]}e^{if\tau}d\tau d\Gamma \tag{5.29}$$

where $y(\tau)$ is the position of the noise source at time τ, Γ is the surface of the noise source and n is the unit normal vector of noise source. Figure 5.3 shows the decomposition of a centrifugal fan volute for implementation of the thin-shaped BEM. The surface of the

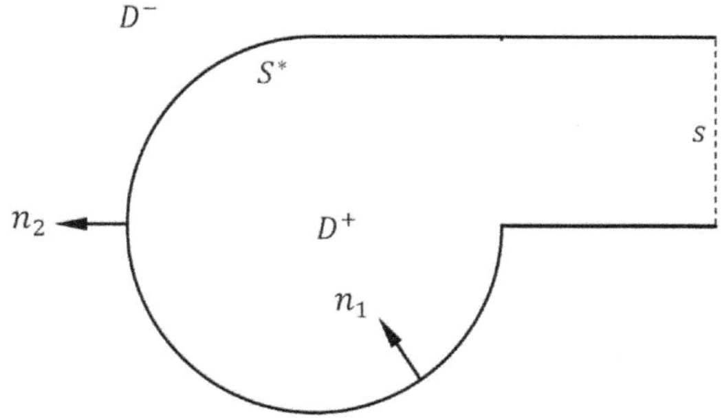

Figure 5.3 Boundary of a centrifugal fan for implementation of the thin-body boundary element method.

volute is denoted by S^* and an imaginary surface s is considered to divide the domain into the exterior subdomain D^- and the interior subdomain D^+. The sound pressure in interior and exterior subdomains is denoted by p^+ and p^-, respectively.

Now Eqn (5.28) for the interior and the exterior subdomains is rewritten as follows:

$$C^+(x)p'^+(x,f) = p'_I(x,f) + \int_{S^*+s} \left[G(x,y,f)\frac{\partial p'^+(y,f)}{\partial n_1(y)} - \frac{\partial G(x,y,f)}{\partial n_1(y)} p'^+(y,f) \right] \mathrm{d}S(y) \tag{5.30}$$

$$C^-(x)p'^-(x,f) = p'_I(x,f) + \int_{S^*+s} \left[G(x,y,f)\frac{\partial p'^-(y,f)}{\partial n_2(y)} - \frac{\partial G(x,y,f)}{\partial n_2(y)} p'^-(y,f) \right] \mathrm{d}S(y) \tag{5.31}$$

where

$$C^+(x) = \begin{cases} 0, & x \in D^- \\ 1, & x \in D^+ \\ 1/2, & x \in S^* \end{cases} \tag{5.32}$$

$$C^-(x) = \begin{cases} 0, & x \in D^+ \\ 1, & x \in D^- \\ 1/2, & x \in S^* \end{cases} \tag{5.33}$$

After manipulation, the following equations are obtained:

$$\frac{1}{2}\left[p'^+(x,f) + p'^-(x,f)\right] = p'_I(x,f) - \int_{S^*} \left[\frac{\partial G(x,y,f)}{\partial n(y)} \left[p'^+(y,f) - p'^-(y,f)\right] \right] \mathrm{d}S(y), x \in S^* \tag{5.34}$$

$$\frac{\partial p'_I(x,f)}{\partial n(x)} - \int_{S^*} \left[\frac{\partial G(x,y,f)}{\partial n(x)\partial n(y)} \left[p'^+(y,f) - p'^-(y,f)\right] \right] \mathrm{d}S(y) = 0, \quad x \in S^* \tag{5.35}$$

where

$$\frac{\partial}{\partial n(y)} = \frac{\partial}{\partial n_1(y)} = \frac{\partial}{\partial n_2(y)} \tag{5.36}$$

The problem of sound-field scattering by the volute of the fan can be simulated after solving Eqns (5.34) and (5.35) to calculate the sound-pressure jump, $p'^{+}(x,f) - p'^{-}(x,f)$, on the wall of the volute first, and then to evaluate the sound pressure at any field point. Equation (5.35) is initially solved using the BEM, which yields the sound-pressure jump on the volute surface. Equation (5.34) is then solved to calculate the total sound pressure $(p'^{+}(x,f) + p'^{-}(x,f))$ in the acoustic domain. There is a vast potential for research on this basis and many groups round the globe are investigating it.

5.4 Effect of fan components on sound generation

The concepts of sound generation of centrifugal fans, noise types and their sources, and numerical simulation of flow-induced sound in fans are introduced in the previous section. In this section, results of some experimental and numerical works on forward-curved centrifugal fans are presented, and the effects of geometrical modifications on performance and sound generation are examined. The utilized numerical and experimental tools are introduced in Chapter 1.

5.4.1 Inlet

It was mentioned in Chapter 4 that a separation zone forms behind the inlet that occupies about one-third of the rotor width. The frequency spectrum of the rotor exit radial velocity in this region shows that the BPF is absent, while the turbulent intensity is high.

Bell-mouth inlets are common in centrifugal fans, which can be installed inward or outward. Figure 5.4 illustrates head coefficient and SPL of three types of fans: two fans with inward inlets and one fan without inlet nozzle. An inward inlet with larger diameter than the rotor inner diameter offers both higher heads and volumetric flow rates (Figure 5.4(a)). The reason for a smaller head coefficient of inlet with smaller diameter than the impeller is that a part of the rotor inlet is not subject to fluid flow and the blade suction is ineffective there. It is observed from Figure 5.4(b) that these fans have similar SPL for the same flow coefficient which of course enhances as the flow rate increases. This shows that the noise source is not the areas close to the inlet, but it is near the backplate where a larger portion of the flow exits the rotor.

There is an annular clearance between the inlet nozzle and rotor when the inlet nozzle is somewhat smaller than the rotor inner diameter. The higher static pressure in the volute as compared to that of the inlet generates a relatively fast flow through the ring gap, which helps the main flow to turn from axial into radial. The annular clearance size is found to be effective in performance and noise. Figure 5.5 presents variation of total head and SPL of squirrel-cage fans with the size of annular clearance between rotor and inlet nozzle. The best performance and SPL are associated with the smallest clearance.

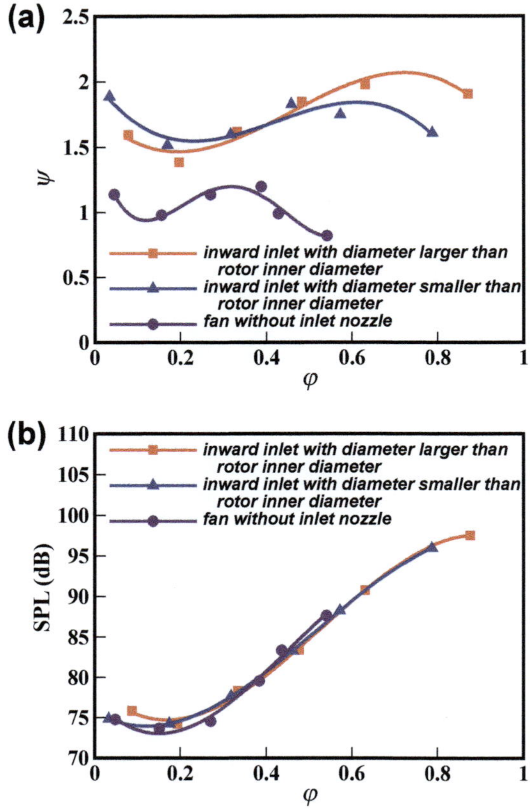

Figure 5.4 Comparison between head coefficient (a) and sound-power level (b) of the fans with inward inlets of different diameters and a fan with no inlet nozzle.

5.4.2 Volute

The role of the volute casing in converting kinetic energy of the rotor outlet flow to static pressure is very important in squirrel-cage fans since the degree of interaction is small as compared to other centrifugal fans. It was mentioned in Chapter 4 that the structure of the flow inside the volute is complicated and involves separated, inverse and vortical flow zones. Interaction of the three-dimensional and unsteady flow with the volute (particularly near the cut-off) is a main source of sound in this fan.

Figure 5.6 shows the frequency spectrum of nodal pressure fluctuation on the volute mid-plane of a squirrel-cage fan. The results are associated with a point close to the cut-off and a point on $\theta = 240°$. The figure shows domination of an amplitude at BPF on the point close to the cut-off, which is not the case for the point on $\theta = 240°$. This is because outlet flow is not uniform around the rotor. The radial distance between the rotor and the volute on $\theta = 240°$ is greater than that at the cut-off. Increasing of rotor–volute distance weakens the rotor–volute interaction and causes the amplitude at BPF to be barely noticeable and masked by broadband level.

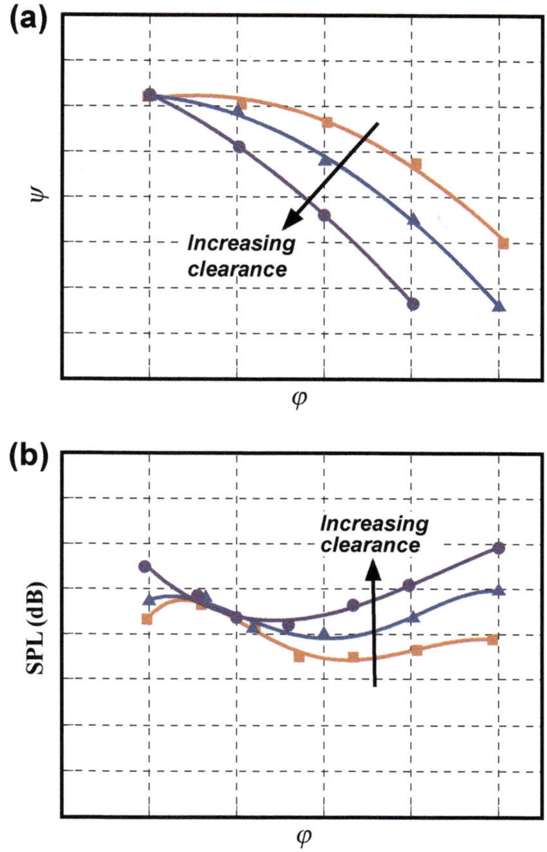

Figure 5.5 Head coefficient (a) and sound-power level (b) of the fans with different rotor–inlet annular clearance.

Circumferential distribution of pressure fluctuation at BPF around the volute casing is shown in Figure 5.7. The highest values of pressure fluctuation are on the cut-off and its neighbourhood (about $\theta = 60°$), which indicates that the cut-off is the main zone of unsteadiness and interaction with the rotor outflow. Away from the cut-off, the pressure fluctuation amplitude decreases and reaches the minimum value around $\theta = 300°$.

The frequency spectrum of sound generated by a squirrel-cage fan in the outlet duct is shown in Figure 5.8 that also shows a dominant peak at BPF. This spectrum is expected to contain wall pressure fluctuations (Figure 5.6), but still the BPF is dominant.

Centrifugal fans are surrounded by a volute, and therefore the generated sound does not propagate in free space. The sound is scattered by the volute with an asymmetric geometry, and therefore the sound far field of the fan also has a nonsymmetric pattern. Results of a numerical study on sound generation of a centrifugal fan based on the thin-body BEM are shown in Figure 5.9. This figure shows the volute-scattering effect on nonsymmetrical sound radiation pattern of the fan. The volute casing causes higher SPL in the frontal zones of the outlet duct than that behind the fan.

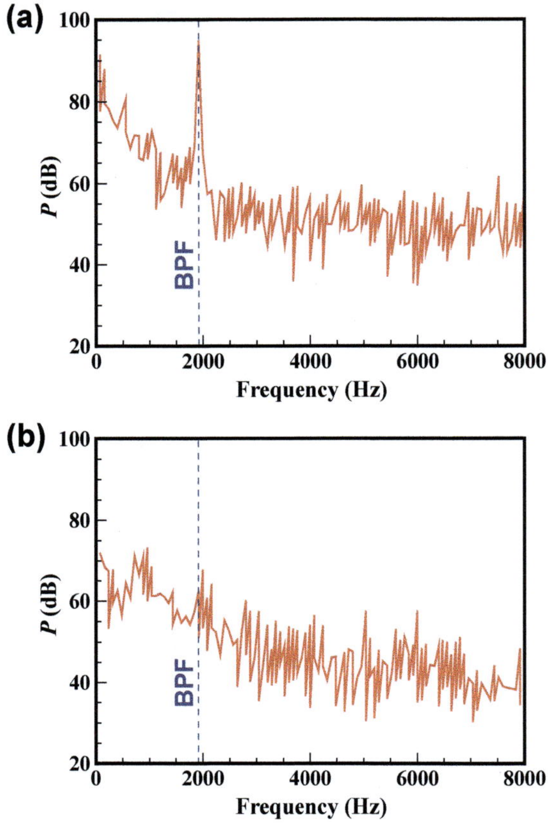

Figure 5.6 Spectrum of pressure fluctuation on a point close to the cut-off (a) and another point at $\theta = 240°$ (b) along the volute mid-plane. The blade passing frequency (BPF) is shown in the dashed line.

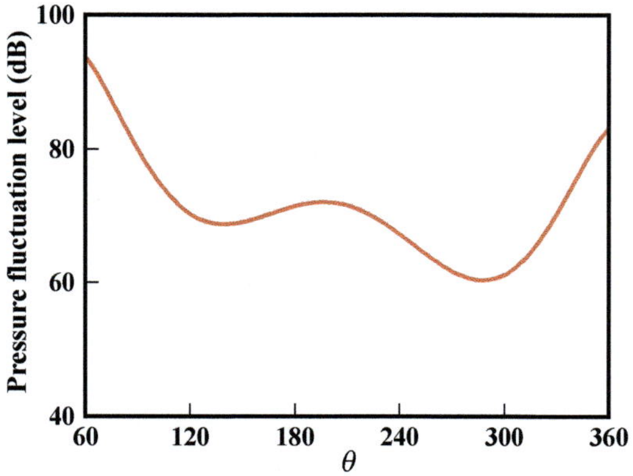

Figure 5.7 Pressure fluctuation level around the volute at blade passing frequency.

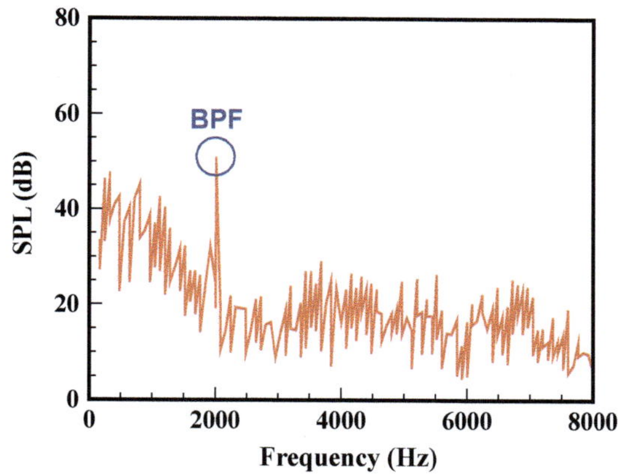

Figure 5.8 Frequency spectrum of sound generated by a squirrel-cage fan at the outlet duct. BPF, blade passing frequency.

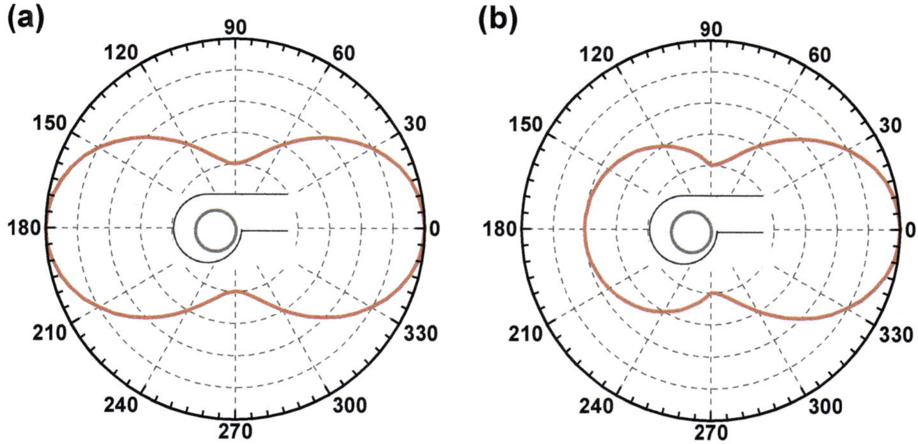

Figure 5.9 The sound radiation directivity pattern of a centrifugal fan: (a) without consideration of the volute effects and (b) with consideration of the volute-scattering effects.

5.4.3 Cut-off

The radial velocity profile exhibits sharp minima and maxima in regions very close to the impeller tip due to blade jet and wake flow. The cut-off is placed in this region and experiences strong pressure fluctuations which in turn results in an effective sound generation at the BPF and its harmonics. After enlarging this clearance, the maxima and minima of the radial velocity profile at the cut-off surface become smoother that diminishes the amplitudes of the pressure fluctuations at the cut-off.

Figure 5.10 shows the trend for noise reduction achieved by increasing the cut-off clearance. The difference in BPF level is plotted versus the clearance, δ, normalized by the impeller diameter, D, which decreases by increasing this clearance.

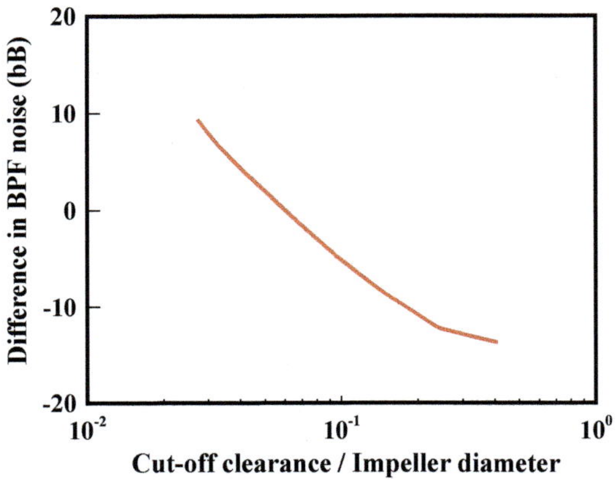

Figure 5.10 Effect of enlarging impeller–cut-off clearance on the difference in blade passing frequency noise level.

Effect of cut-off radius on BPF sound is much less than that of the cut-off clearance. Experiments on three fans with different values of clearance show that increasing the cut-off radius from $r/D = 0.005$ to $r/D = 0.1$ reduces the maximum sound about 6 dB.

The geometrical modification of the cut-off may affect sound and performance of the fan. Figure 5.11 shows two possible geometrical modification of the cut-off with installation of cubic or wedge-shaped barriers on the cut-off. The rotor–barrier tip distance increases by enlarging the barrier, while the height of the outlet channel decreases. Figure 5.12 presents SPL measurements for the original and the modified volutes. The experiments show that SPL of the fans with modified cut-off increases by flow rate, while it marginally increases for the original fan. Also the SPL at high rotational speeds are greater than those at moderate rotational speeds.

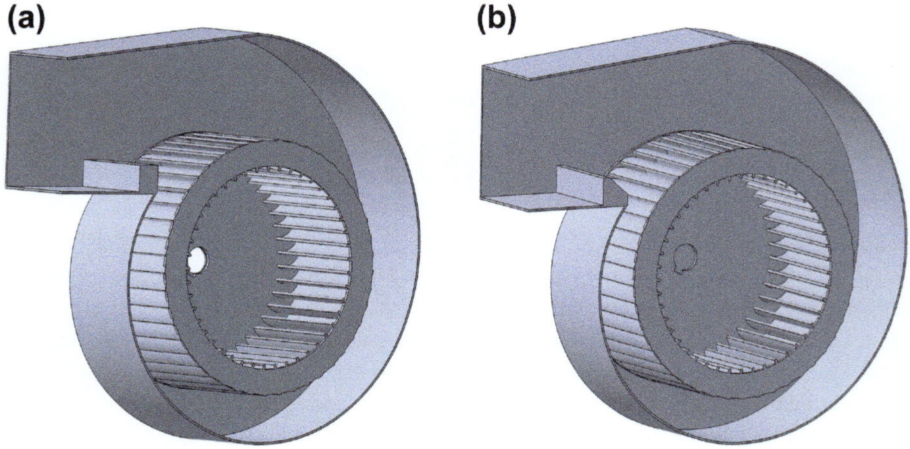

Figure 5.11 Modification of the cut-off by installing a cubic (a) or a wedge-shaped (b) barrier above the cut-off.

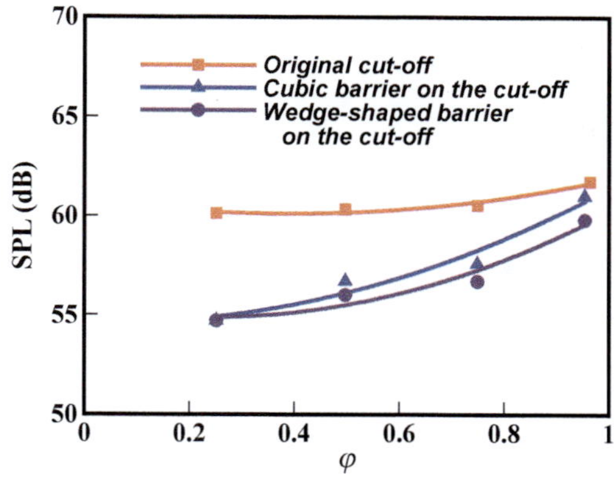

Figure 5.12 Effect of cut-off modifications on the total sound-pressure level (SPL).

According to the experiments, a reduction of the fan outlet height reduces the sound generated by the fan, especially for low and medium flow rates (Figure 5.12).

Another type of cut-off configuration is the inclined type. Figure 5.13 shows a schematic of this type of cut-off where the radius varies from the inlet side to the volute back plate.

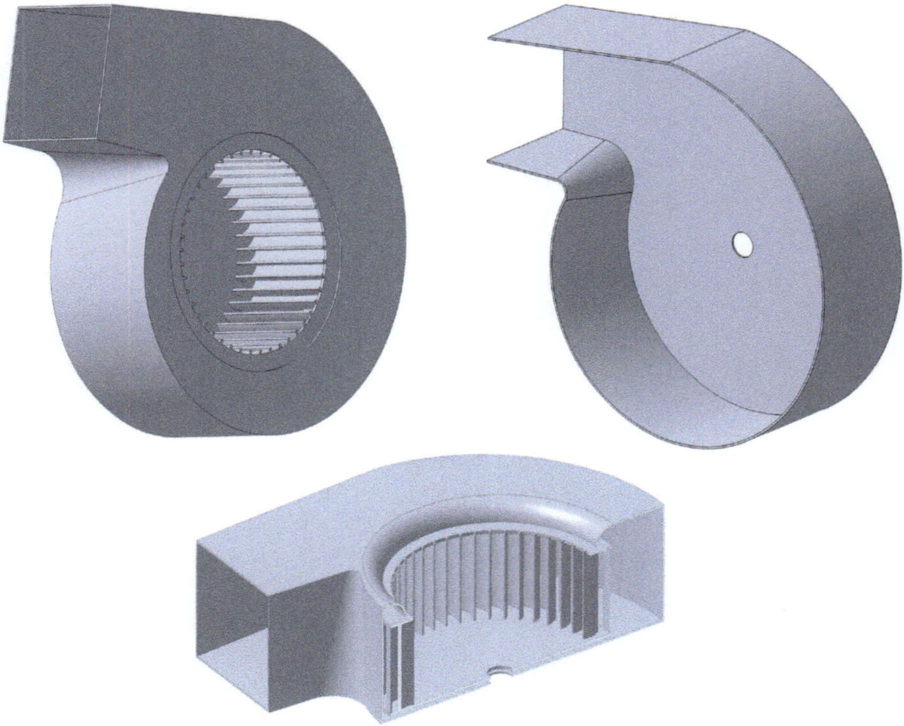

Figure 5.13 Schematic of the inclined cut-off.

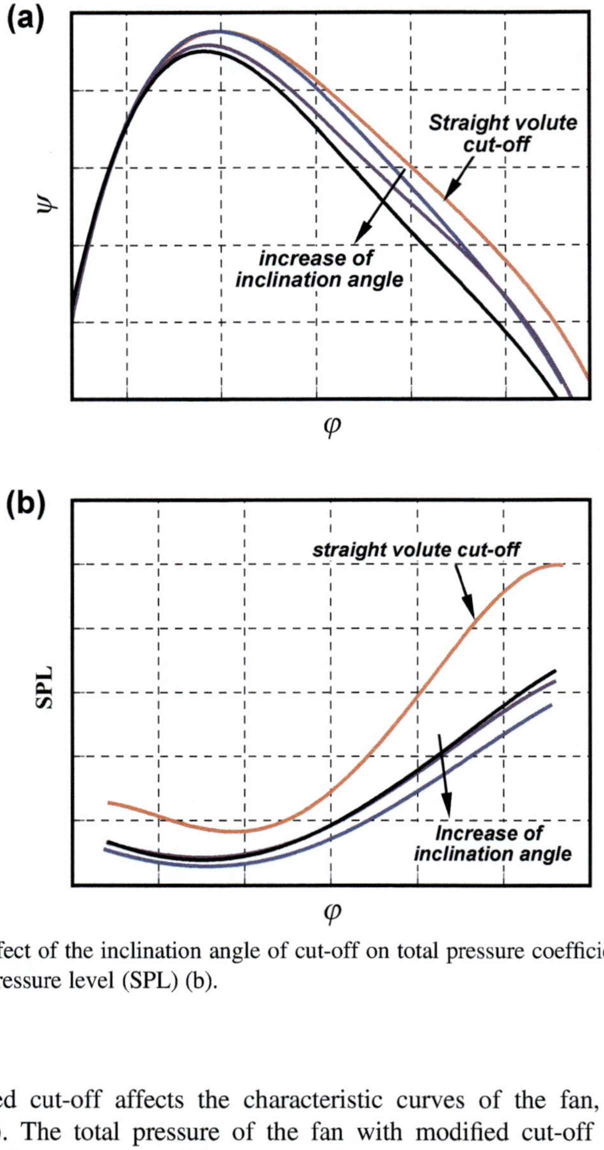

Figure 5.14 Effect of the inclination angle of cut-off on total pressure coefficient (a) and the overall sound-pressure level (SPL) (b).

The inclined cut-off affects the characteristic curves of the fan, as shown in Figure 5.14(a). The total pressure of the fan with modified cut-off is lower than that of the original fan at moderate and high flow rates. It is expected that for larger inclination angles, the distance between the rotor and cut-off increases too, which reduces the intensity of pressure fluctuations on the cut-off and consequently the sound at BPF.

As shown in Figure 5.14(b), the modified inclined cut-offs affect the total SPL of the fan. All the three modified cut-offs decrease the noise as compared with the original fan. The effect of different cut-offs is not apparent at low flow rates, while at high flow rates the cut-off with the highest inclination angle results in minimum SPL.

5.4.4 Rotor

The width-to-inner diameter ratio of the impeller is generally selected in the design stage to be large enough to increase flow rate and pressure rise. However, as is stated in Chapter 4, wider impeller and the ensuing wider separation zone at the impeller shroud side deteriorate performance. Although there is a change in performance, other experiments show the negligible effect of impeller width-to-diameter ratio on the sound generation of fans.

The position of the rotor relative to the inlet nozzle and volute backplate (Figure 1.1) may affect the performance and sound generation of fan. Figure 5.15 shows that axial clearance between the inlet nozzle and shroud plate has almost no effect on the head coefficient and aerodynamic sound. This phenomenon has two reasons. First is that the degree of reaction in squirrel-cage fans is small. Consequently, the static pressure rise at the exit of the impeller is too small to affect leakage from the clearance. The second reason is the separation at the impeller entrance regardless of the clearance geometry. Even if the leakage flow occurs because of the clearance, the blade passage flow is hardly influenced by this leakage flow.

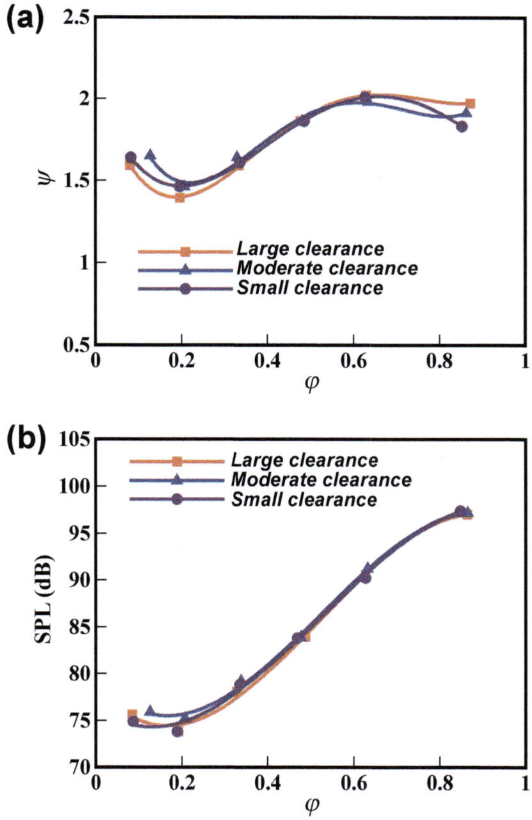

Figure 5.15 Head coefficient (a) and sound-power level (SPL) (b) of the fans with different impeller-inlet clearance.

The effect of hub–volute clearance (the distance between the rotor backplate and the volute in Figure 1.1) on total pressure and total SPL of a forward-curved centrifugal fan is shown in Figure 5.16. As the flow coefficient increases, larger hub–volute clearance enhances total pressure rise in the fan. Variation of total SPL with hub–volute clearance in Figure 5.16(b) shows that at high flow rates, which are the usual operating conditions, a good noise reduction is obtained for a larger hub–volute clearance, while at moderate flow rates, total noise of all modified fans is higher than that of the original fan. The fan with the largest hub–volute clearance has therefore the best enhancement in performance and reduction in noise. No performance limit is measured for this parameter.

The shroud can affect fan performance through varying the size of the separation zone behind the inlet as is already shown in Chapter 3. The shroud can be replaced with two rings that support the blades. Figure 5.17 shows that replacing the shroud with two rings has no considerable influence on the SPL of the fan, but its head coefficient decreases as much as 5%.

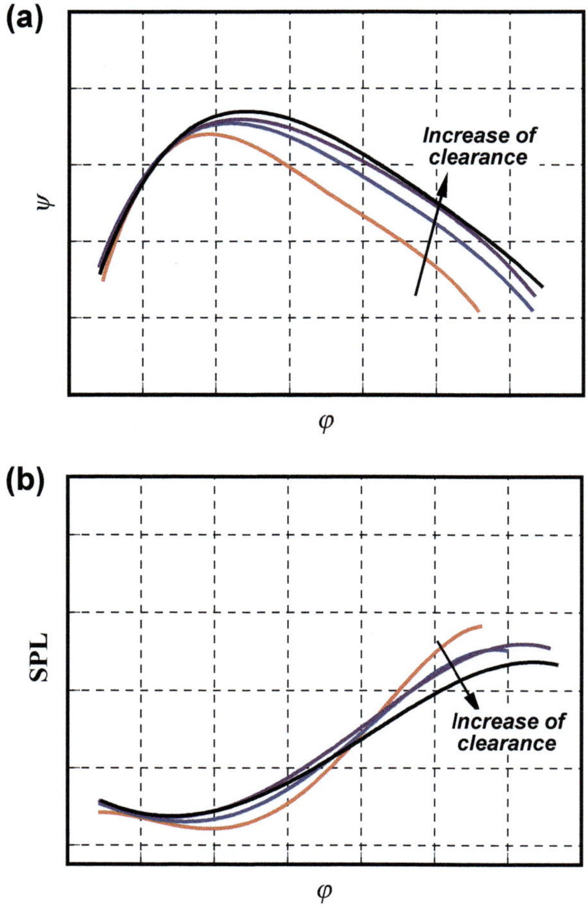

Figure 5.16 Effect of hub–volute clearance on head coefficient (a) and the overall sound-pressure level (SPL) (b).

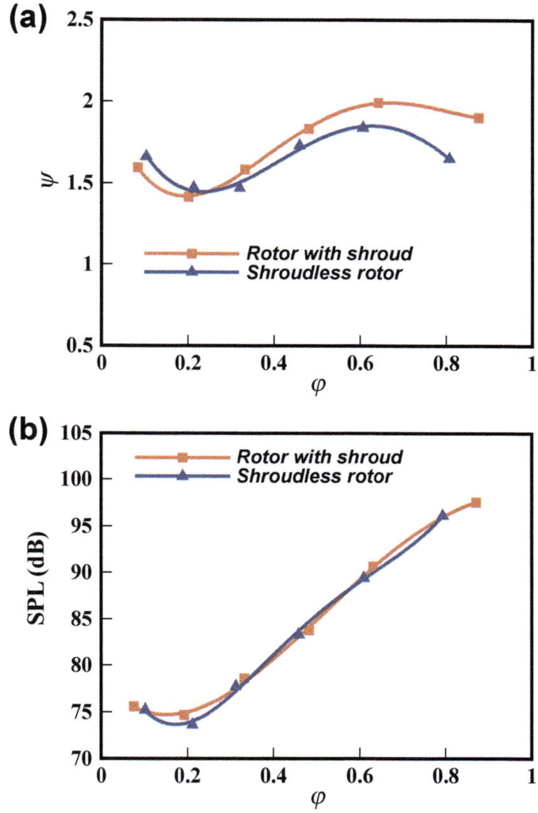

Figure 5.17 Effect of existence of shroud on head coefficient (a) and sound-power level (SWL) (b) of squirrel-cage fans.

Blade setting on the rotor (Figure 5.18) is an important parameter that influences fan performance and sound generation. The blade outlet angle changes with the blade-setting angle. Subsequently, the impeller-outlet relative velocity, turbulence and the sound level increase due to the flow separation at the blade suction surfaces. It is known that an optimal value of the blade setting angle minimizes the fan noise. The optimal value is measured to be 26°.

Parallel cut-off and blade trailing edge are commonplace in design and production of squirrel-cage fans where pressure pulses from the blades on the cut-off are in phase. In case of an inclined cut-off or inclined blades or both, a phase shift of the pressure wave on the cut-off is experienced. This phenomenon results in local cancellation of pressure fluctuations and consequently reduces SPL. The inclined cut-off was considered in the previous subsection. Rotor lean angle was also introduced in Chapter 3. The experimental investigations show that inclined blades of the forward-curved centrifugal fans reduce 12 dB in BPF noise for constant rotor speed. It should be noted that the inclined cut-offs are easier to produce and more effective in noise reduction.

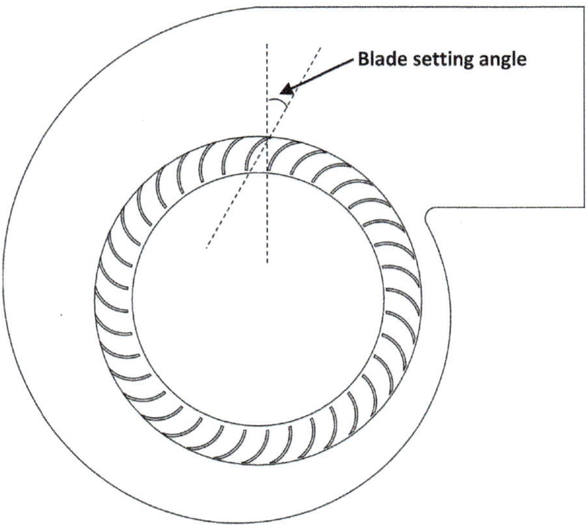

Figure 5.18 Definition of blade setting angle on the rotor.

When blades with unequal spaces are mounted on the impeller, the energy of BPF noise spreads over a wider frequency band. Experimental studies show that although the BPF noise might reduce, the total noise generated in this type of fan remains unchanged. Moreover, it is found from experiments that at certain operating conditions, the fan with unequally spaced blades generates side band noise that is higher than the BPF noise.

Double-inlet squirrel-cage fans have two separate half rotors. Such a fan is shown in Figure 5.19. If the blades of one-half lie halfway between the blades

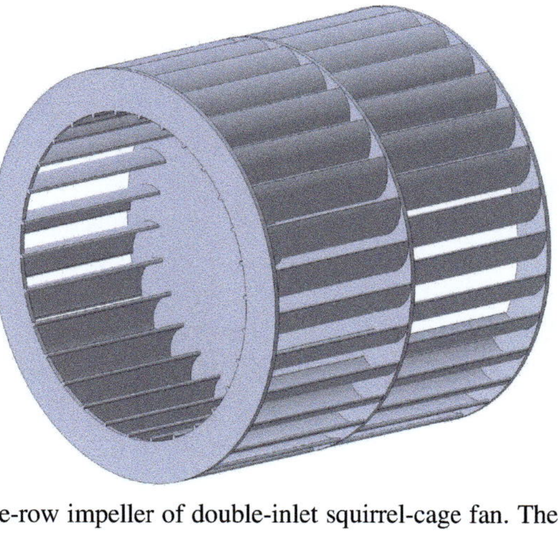

Figure 5.19 A double-row impeller of double-inlet squirrel-cage fan. The blades of the left-hand half-rotor are shifted to the halfway between the blades of the right-hand half-rotor.

of the other half, the produced pressure jets along the cut-off are out of phase, which results in the reduction of BPF noise up to 8–10 dB. This method is easily applicable to small squirrel-cage fans but in large fans with welded units, such asymmetric construction causes thermal stress problems and deformations in the impeller backplate.

An old technique for noise control in forward-curved fans is the installation of transition meshes at the impeller leading and trailing edges. The mesh at the leading edge of the blades forms small-scale turbulent flow downstream the mesh and turbulent boundary layer close to the blades, which together shift the flow separation further downstream. The mesh at the trailing edge of the blades makes the outlet jet and wake velocity profile smoother and reduces the turbulence interactions.

Figure 5.20 illustrates the effect of fine and coarse meshes on reduction of sound generated by forward-curved fans. The fine mesh reduces the noise in the frequency range of 100–10 kHz by 4–9 dB. The drawback is a 6% loss in the pressure head and 8% drop of efficiency. A significantly lower noise in the frequency range of 100–2000 Hz is obtained with a coarse mesh, while the high-frequency noise is intensified. Moreover, loss in pressure head and efficiency due to the coarse mesh is higher than those by a fine mesh.

The idea of half-cone rotors was introduced in Chapter 3 (Figure 3.1). These geometries aim to align the impeller with the inlet flow. The numerical results on noise generation show that the impeller with the negative angle of $-10°$ has the highest noise, while the impeller with the angle of $+10°$ has the lowest noise. Generated noise by cylindrical impellers is moderate in comparison. It was shown in Chapter 4 that the outflow from the rotor is near the backplate. It is also found in this chapter that the BPF noise in squirrel-cage fans decreases by increase in cut-off clearance. Positive half-cone rotors have a larger distance between the rotor and the cut-off closer to the backplate, where the dominant rotor outlet flow occurs. Therefore, the BPF noise of these rotors and consequently their total SPL is lower than the cylindrical and negative half-cone rotors.

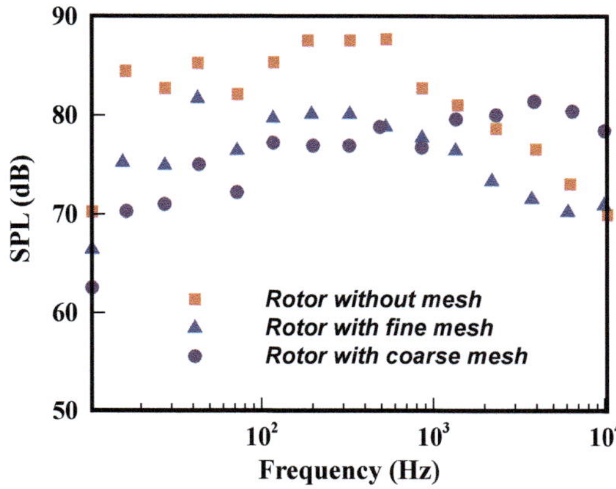

Figure 5.20 Sound generation in centrifugal rotors equipped with different scales of mesh.

5.5 Sound generation of double-outlet squirrel-cage fans

This type of fan, introduced in Chapter 4, can be used in multiple air-delivery zone air-conditioning systems. It seems that since two noise sources (cut-offs) exist in this type of fan, they can interact to dampen pressure fluctuations or resonate to increase it. This section discusses the available knowledge on efficiency and noise generation of double-outlet squirrel-cage fans. Experiences on two double-outlet squirrel-cage fans with different values of cut-off clearance are presented.

Figure 5.21 shows variations of efficiency and pressure coefficient of double-outlet squirrel-cage fans versus flow rate. There is no standard that defines how to combine pressure rise and flow rates of the two outlets. The efficiency is calculated in each operating conditions for the total flow rate of both upper and lower outlets, while the pressure coefficient is obtained for each outlet individually. The fan with a smaller cut-off clearance has a higher efficiency and total pressure for a constant flow rate than the fan

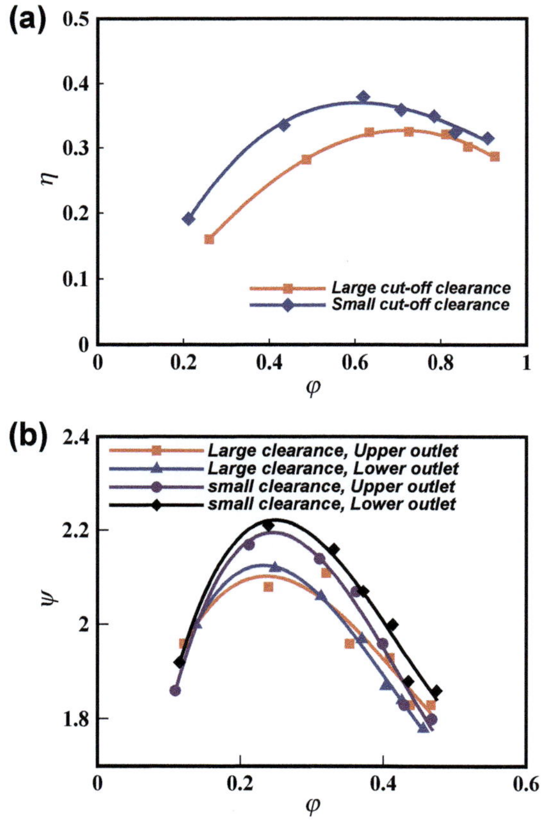

Figure 5.21 Variation of total efficiency (a) and total pressure coefficient (b) versus flow rate for two double-outlet fans with different cut-off clearances.

with larger cut-off clearance. This agrees with observations in single-outlet fans that the decrease in cut-off clearance enhances efficiency. Furthermore, at lower flow rates, the difference between efficiencies of the two fans is notable and for larger flow rates, especially for the values greater than BEP, the difference between head coefficients is negligible.

Figure 5.22 depicts the frequency spectra of two double-outlet fans with different cut-off clearance at a low flow rate. Measurements are in the outlet ducts. A peak corresponding to rotational frequency, RF (=12 Hz) is observed for both fans. There is a peak at the BPF (=530 Hz) in all cases. The figure also shows broadband noise with decreasing amplitude from low frequencies up to 2000 Hz. Moreover, some peaks in frequencies around the BPF are observed for both double-outlet fans. These peaks occur at BPF + RF and BPF − RF. It is suggested in the literature that these noise components could correspond to resonance and vibration of fan components, especially the casing, and are not aerodynamic phenomena.

Figure 5.23 illustrates variation of total SPL in the outlet ducts versus flow-rate coefficient for two double-outlet and one single-outlet fans. In the case of double-outlet

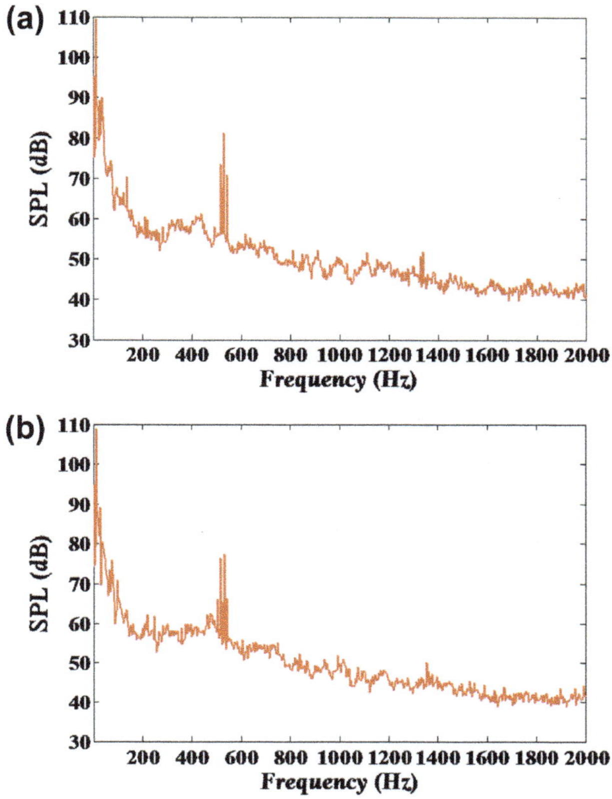

Figure 5.22 Sound-pressure level (SPL) frequency spectrum of double-outlet squirrel-cage fan with larger (a) and smaller (b) cut-off clearance.

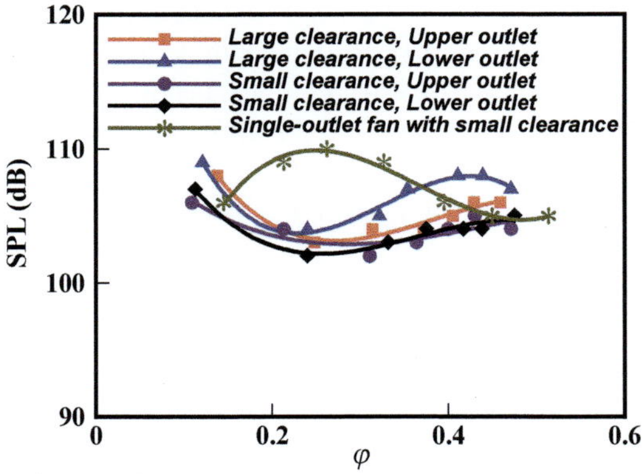

Figure 5.23 Variation of total sound-pressure level (SPL) versus flow coefficient for two double-outlet and one single-outlet fans.

fans, maximum SPL occurs at minimum flow rate, and then a significant decrease in SPL is observed by increasing the flow rate. When these fans operate close to the BEP, minimum or moderate SPL is generated. A high value of SPL is found for high flow rates, but this is not greater than the SPL at the lowest flow rate. Results for a single-outlet fan with small cut-off clearance are also presented in the figure for comparison. A different behaviour is found for single-outlet fan, such that the minimum SPL occurs in lowest and highest flow rates and the highest SPL is experienced at moderate flow rates. This behaviour is quite different with that of double-outlet fans that have the optimal SPL at BEP.

Figure 5.24 depicts SPL at BPF versus flow rate for two double-outlet and one single-outlet fans. The BPF noise of the double-outlet fan with large clearance increases with flow rate until $\varphi \approx 0.4$, where the maximum amplitude of 82 dB for upper and lower outlets is evident. The BPF noise decreases with further increase in flow rate. The double-outlet fan with small clearance is somehow different, since increasing

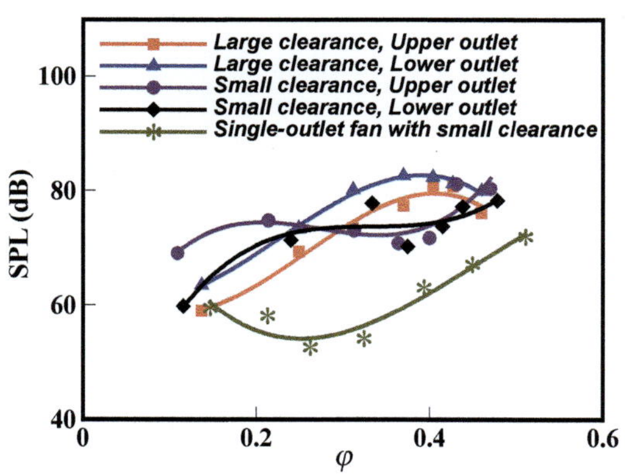

Figure 5.24 Variation of noise at blade passing frequency versus flow coefficient for two double-outlet and one single-outlet fans.

of the BPF noise is from the minimum flow rate until $\varphi \approx 0.3$. Above this flow rate, the SPL decreases slightly and further increase in flow rate motivates the BPF noise to increase to maximum amplitude for both upper and lower outlets at maximum flow rate. This might be specific to this fan, but its occurrence is important since it can bring about a noise drop in the operating range of these fans. A comparison between BPF noise of double-outlet fans shows that the fan with smaller cut-off clearance generates lower tonal noise for approximately all the operating range. Previous works about noise generation in single-outlet squirrel-cage fans state that when the cut-off clearance increases, more uniform flow interacts with the cut-off and a lower BPF noise is generated. The experiments on a double-outlet fan make a different observation. It complies with the idea that properly tuning of the two cut-offs can dampen the generated noise by each of them. Frequencies switch power, and therefore noise is distributed between different frequencies. Further analysis of this phenomenon is still required which is beyond the scope of this book.

The overall examination of Figure 5.24 shows that the BPF noise of double-outlet fans is higher than that of single-outlet fan. This is because of the existence of two noise sources (cut-off) in double-outlet fans. The flow rates in each of the channels of double-outlet fans are set to be equal to the flow rate of single-outlet fan. Since the SPL at BPF is proportional to the flow rate, it can be concluded that the noise generated by interaction of the rotor outflow and the cut-off is similar for single- and double-outlet fans. The BPF noise generated in each cut-off is propagated in both outlet channels of the double-outlet fan and consequently the BPF noise in each channel is composed of two distinct noise sources.

The BPF noise of the double-outlet fan with smaller cut-off clearance decreases for flow rates around $\varphi = 0.4$. On the other hand, noise in frequencies around BPF (BPF $\pm$ 12 Hz) is observed in the frequency spectrum of this fan (Figure 5.22). Figure 5.25 depicts variations of SPL at BPF, BPF $-$ 12 Hz and BPF $+$ 12 Hz versus flow rate for a double-outlet fan with small cut-off clearance in order to clarify the different behaviour of BPF noise of this fan. The results show that at flow rates lower

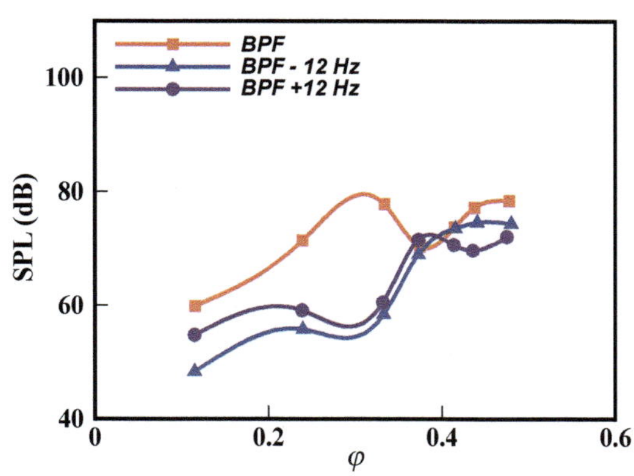

Figure 5.25 Variations of noise at blade passing frequency (BPF) versus flow coefficient for a double-outlet fan with small cut-off clearance.

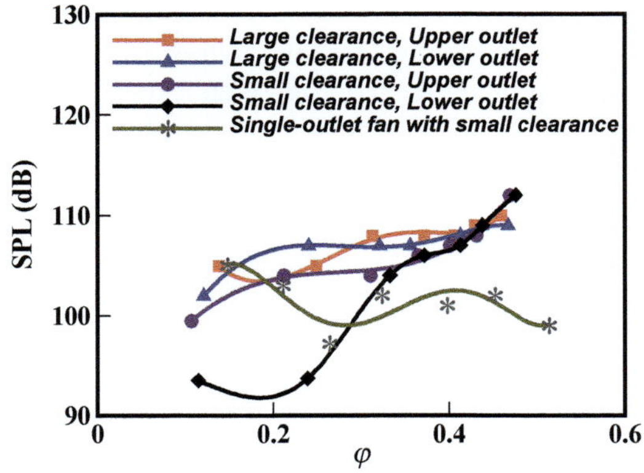

Figure 5.26 Variation of sound-pressure level of single- and double-outlet fans at rotation frequency versus flow coefficient.

than 0.35, the corresponding SPLs to BPF ± 12 Hz are significantly lower than that of BPF. In the normal operating points the SPL at BPF decreases dramatically, but a significant increase in noise at BPF ± 12 Hz is evident. The SPL at these frequencies is higher than that of BPF. It shows that there is a mechanism that shifts the acoustic energy from BPF to BPF ± 12 Hz. It should be noted that this phenomenon is not observed for other fans.

Figure 5.26 compares the SPL spectra of two double-outlet and one single-outlet fans at the RF. It shows that in double-outlet fans the SPL slightly increases with flow rate, while in single-outlet fan the values remain at about 100 dB. The noise in such a frequency is related to mechanical sources. It is worth noting that this frequency can be outside the hearing range for human beings, and therefore it is more important to keep it out of the hearing range rather than dampening it.

5.6 Final remarks

The effect of different geometries on fan noise has been discussed in this chapter. The inlet makes little change on noise while the rotor and volute are important as far as the proximity of the blades to the cut-off occurs. The frequency content of fan noise can change, and if properly utilized, can reduce the overall noise.

Further reading

Ballesteros-Tajadura, R., Velarde-Suárez, S., & Hurtado-Cruz, J. P. (2008). Noise prediction of a centrifugal fan: numerical results and experimental validation. *Journal of Fluids Engineering, 130*, 1–12.

Ballesteros-Tajadura, R., Velarde-Suárez, S., Hurtado-Cruz, J. P., & Santolaria-Morros, C. (2006). Numerical calculation of pressure fluctuations in the volute of a centrifugal fan. *Journal of Fluids Engineering, 128*, 359–369.

Bayomi, N. N., Abdel Hafiz, A., & Osman, A. M. (2006). Effect of inlet straighteners on centrifugal fan performance. *Energy Conversion and Management, 47*, 3307–3318.

Chunxi, L., Song Ling, W., & Yakui, J. (2011). The performance of a centrifugal fan with enlarged impeller. *Energy Conversion and Management, 52*, 2902–2910.

Cory, W. T. W. (1992). Acoustic similarity laws for the prediction of industrial fan sound level. In *Proceeding of the SFA symposium on fan noise, Senlis, France.*

Curle, N. (1955). The influence of solid boundaries upon aerodynamic sound. In *Proceedings of the Royal Society of London; Series A: Mathematical and Physical Sciences, 231* (pp. 505–514).

Datong, Q., Yijun, M., Xiaoliang, L., & Minjian, Y. (2008). Experimental study on the noise reduction of an industrial forward-curved blades centrifugal fan. *Applied Acoustics, 70*, 1041–1050.

Embleton, T. F. W. (1963). Experimental study of noise reduction in centrifugal blowers. *The Journal of the Acoustical Society of America, 35*, 700–705.

Farassat, F., & Brentner, K. S. (2003). Modeling aerodynamically generated sound of helicopter rotors. *Proceedings of Royal Society of USA, 23681*, 94–96.

Ffowcs-Williams, J. E., & Hawkings, D. L. (1969). Sound radiation from turbulence and surfaces in arbitrary motion. *Proceedings of Royal Society of London*, 321–342.

Hessami Azizi, M. H., Montazerin, N., & Damangir, A. (2005). A combined noise-laser Doppler velocimetry study of the squirrel-cage fan. *Iranian Journal of Mechanical Engineering, 6*, 5–17.

Hosseini, S. A. (2012). Experimental study of noise interaction between two outlets of squirrel-cage fan, MSc Thesis, Amirkabir University of Technology. (in Persian).

Hu, B., OuYang, H., Wu, Y., Jin, G., Qiang, X., & Du, Z. (2013). Numerical prediction of the interaction noise radiated from an axial fan. *Applied Acoustics, 74*, 544–552.

Lighthill, M. J. (1952). On the sound generation aero-dynamically: I. General theory. *Proceedings of Royal Society of London; Series A: Mathematical and Physical Science, 211*, 564–587.

Lighthill, M. J. (1954). On the sound generated aerodynamically: II. Turbulence as a source of sound. *Proceedings of Royal Society of London; Series A: Mathematical and Physical Science, 222*, 1–22.

Liu, Q., Qi, D., & Mao, Y. (2006). Numerical calculation of centrifugal fan noise. *Proceedings of the Institution of Mechanical Engineers, Part C: Journal of Mechanical Engineering Science, 220*, 1167–1177.

Liu, Q., Qi, D., & Tang, H. (2007). Computation of aerodynamic noise of centrifugal fan using large eddy simulation approach, acoustic analogy, and vortex sound theory. *Proceedings of the Institution of Mechanical Engineers, Part C: Journal of Mechanical Engineering Science, 221*, 1321–1332.

Lyons, L. A., & Platter, S. (1963). Effect of cut-off configuration on pure tones generated by small centrifugal blowers. *The Journal of the Acoustical Society of America, 35*, 1455–1456.

Mao, Y., & Qi, D. (2009). Computation of rotating blade noise scattered by a centrifugal volute. *Proceedings of the Institution of Mechanical Engineers, Part A: Journal of Power and Energy, 223*, 965–972.

Mao, Y., Qi, D., Liu, X., & Tang, H. (2008). Numerical prediction of aerodynamic tonal noise radiated from a centrifugal fan. *Proceedings of the Institution of Mechanical Engineers, Part A: Journal of Power and Energy, 222*, 831–842.

Montazerin, N. (2000). A double-outlet squirrel-cage fan. In *5th Biennial conference on engineering systems design and analysis, ASME 2000.*

Neise, W. (1976). Noise reduction in centrifugal fans: a literature survey. *Journal of Sound and Vibration, 45*, 375–403.

Neise, W. (1992). Review of fan noise generation mechanism and control methods. In *Proceeding of the SFA symposium on fan noise, Senlis, France.*

Neise, W. (1982). Review of noise reduction methods for centrifugal fans. *Journal of Engineering for Industry, 104*, 151–161.

Powell, A. (1964). Theory of vortex sound. *Journal of the Acoustical Society of America, 36*, 177–195.

Shetab, M. R. (2011). Investigation of acoustic field of squirrel-cage fan with half-cone rotor, MSc Thesis, Amirkabir University of Technology. (in Persian).

Velarde-Suárez, S., Ballesteros-Tajadura, R., Santolaria-Morros, C., & González-Pérez, J. (2001). Unsteady flow pattern characteristics downstream of a forward-curved blades centrifugal fan. *Journal of Fluids Engineering, 123*, 265–270.

Velarde-Suárez, S., Ballesteros-Tajadura, R., & Hurtado-Cruz, J. P. (2006). A predictive maintenance procedure using pressure and acceleration signals from a centrifugal fan. *Applied Acoustics, 67*, 49–61.

Velarde-Suárez, S., Ballesteros-Tajadura, R., Gonzalez-Perez, J., & Pereiras-Garcia, B. (2006). Numerical simulation of the unsteady flow patterns in a small squirrel-cage fan. In *Proceedings of FEDSM2006 ASME Joint U.S. – European Fluids Engineering Summer Meeting, Miami.*

Velarde-Suárez, S., Ballesteros-Tajadura, R., Hurtado-Cruz, J. P., & Santolaria-Morros, C. (2006). Experimental determination of the tonal noise sources in a centrifugal fan. *Journal of Sound and Vibration, 295*, 781–796.

Velarde-Suárez, S., Ballesteros-Tajadura, R., Santolaria-Morros, C., & Pereiras-García, B. (2008). Reduction of the aerodynamic tonal noise of a forward-curved centrifugal fan by modification of the volute tongue geometry. *Applied Acoustics, 69*, 225–232.

Velarde-Suárez, S., Ballesteros-Tajadura, R., González-Pérez, J., & Pereiras-García, B. (2009). Relationship between volute pressure fluctuation pattern and tonal noise generation in a squirrel-cage fan. *Applied Acoustics, 70*, 1384–1392.

Velarde-Suarez, S., Santolaria-Morros, C., & Ballesteros-Tajadura, R. (1999). Experimental study on the aeroacoustic behavior of a forward-curved blades centrifugal fan. *Journal of Fluids Engineering, 121*, 276–281.

Wang, M., Freund, J. B., & Lele, S. K. (2006). Computational prediction of flow-generated sound. *Annual Review of Fluid Mechanics, 38*, 483–512.

Younsi, M., Bakir, F., Kouidri, S., & Rey, R. (2007). 3D unsteady flow in a centrifugal fan: impeller-volute interaction. *Journal of Computational and Applied Mechanics, 8*, 211–223.

Younsi, M., Bakir, F., Kouidri, S., & Rey, R. (2007). Numerical and experimental study of unsteady flow in a centrifugal fan. *Proceedings of the Institution of Mechanical Engineers, Part A: Journal of Power and Energy, 221*, 1025–1036.

Contribution of jet–wake–volute interactions to flow characteristics and turbulence models

6

6.1 Role of nonintrusive measurements on examination of complicated turbulent flows

Turbomachinery flows are characterized by asymmetric jet–wake interactions and anisotropic flow structures, and these result in a complicated and three-dimensional flow, particularly at near-wall regions. Implementation of intrusive velocimetry techniques such as Pitot tube and hot wire alters the near-probe flow patterns. The vortical turbomachinery flow field is significantly sensitive to these measurement deviations. Advent of nonintrusive optical techniques such as laser Doppler anemometry (LDA) and particle image velocimetry (PIV) considerably enhanced the measurement capability for reliable analysis of flow characteristics, and both temporal and spatial patterns could be investigated. The LDA has stronger potential for temporal analyses due to its high frequency, and PIV is more appropriate for spatial examination of flow structures.

The point-wise LDA measurements in various locations of squirrel-cage fan were analysed in the previous chapters. This chapter discusses the PIV-based patterns in the rotor exit region. The PIV as a spatial or temporal measurement technique plays an important role in obtaining knowledge about flow structures in turbomachinery applications. This is a practical tool at present due to limitations of computational resources in DNS (Direct Numerical Simulation) and numerical challenges for implementation of various turbulence models in such complicated flow fields.

A planar stereoscopic particle image velocimetry (SPIV) extracts all three components of instantaneous velocity in a plane. Overlapping fields of view (FOVs) enlarge the measurement area without sacrificing the resolution. The dimensions of these FOVs are a compromise between capturing fine-scale flow structures and covering sufficient number of jet–wake regions. The area in the upper section of volute (close to the outlet) with 40% axial span with respect to the volute backplate has the most pronounced jet–wake patterns out of the rotor. Recirculations back into the rotor are expected to be minimum at the best efficiency point. Such measurement area and operating condition are selected because of severe jet–wake–volute interactions and their contribution to complexity of turbulent-flow dynamics. Figure 6.1 illustrates the location and dimension of the four FOVs with respect to the rotor. The overlaps consider mutual matching of each pair of FOVs after measurements.

Developments in Turbomachinery Flow.

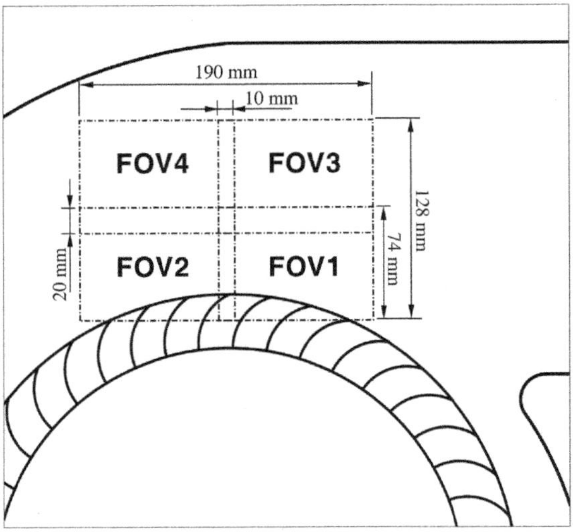

Figure 6.1 Location and dimension of the four fields of view in the rotor exit region.

The three instantaneous velocity components in the measurement plane provide a complete calculation of gradient-less flow quantities such as velocity fluctuations, Reynolds and sub-grid scale (SGS) stresses. However, strain rate tensor, vorticity vector and various turbulent kinetic energy terms such as production, convection, turbulent diffusion and viscous dissipation include out-of-plane velocity gradients that are not calculable from single-plane SPIV data. Consequently, calculation of these quantities requires approximations or assumptions to compensate the missing terms. Despite this drawback, the SPIV data are quite beneficial for turbulent-flow analysis in such complicated geometries because the measured field is realistic and cannot be provided by the simulation-based methods at the present time.

The following sections examine various characteristics of turbulent rotor exit flow as deduced from SPIV. They include jet–wake–volute interactions and their consequences, evaluation of geometrical characteristics of flow structures and assessment of turbulence-modelling challenges in such complicated flow field.

6.2 Jet–wake interactions with the volute flow

Interactions between high-momentum jet streams, low-momentum vortical wake structures and cyclic volute flow in the rotor exit region result in strong shear flows with elevated turbulent intensity. Severe interactions among jet–wake–volute flow structures contribute to energy loss and reduce the performance of each centrifugal turbomachine. The loss is more significant in critical sections of the machine, namely around the cut-off, behind the inlet (recirculation zone) and along the suction side of blades. This is particularly important in off-design condition. Understanding rotor exit flow characteristics and their consequences on fan deficiencies are important for any geometrical/operational optimization.

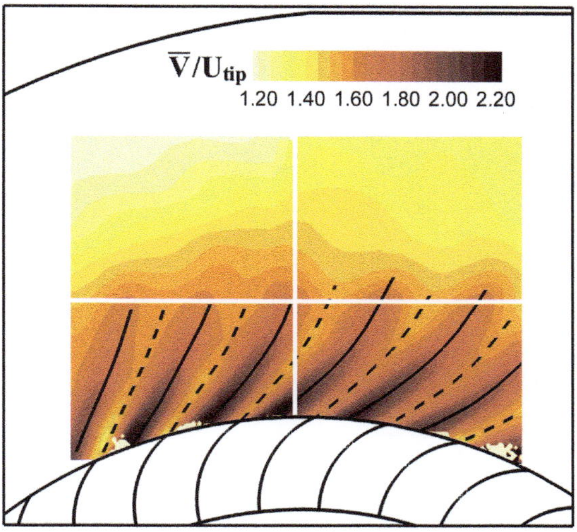

Figure 6.2 Ensemble average of normalized total velocity in the rotor exit region. Jet and wake centrelines are illustrated by solid and dashed curves, respectively.

Figure 6.2 illustrates ensemble average of total velocity in the rotor exit. The velocity is normalized by rotor tangential velocity at the outer radius. The coordinates of local maximums and minimums of velocity for each blade passage are jet and wake centrelines (solid and dashed curves in Figure 6.2), respectively. Jet and wake structures are strong as they exit from the rotor. However, shear interaction and mixing of neighbouring jet–wake structures reduce their intensity and make them less distinguishable. In addition, interaction of these structures with the volute flow (which is formed from the rotor exit jet–wake flow in narrower sections of the volute) generates cyclic mean flow patterns that are shown in Figure 6.2.

The gradient-based quantities such as strain rate and vorticity are superior descriptors of the shear flows. Figure 6.3 shows the normal component of strain rate tensor and its shear component inside x–y (measurement) plane. Positive and negative values of normal strain component correspond to extensive and contractive flow structures, respectively. On the other hand, the opposite signs for shear strain component indicate the two opposing directions of fluid angular deformation.

Figure 6.3(a) for normal x-component deformation shows that the most extensive structures are along the blades pressure side while the most contractive structures occur along the blades suction side. The opposite trend is true for the y-component of normal strain (Figure 6.3(c)). In other words, along the pressure sides of the rotor, the flow accelerates in the x-direction and decelerates in the y-direction, while an opposite behaviour is present along the suction sides. These opposite trends are due to counter-rotating structures along the two sides of each blade. Figure 6.4 presents the contour plot of normalized out-of-plane component of vorticity and confirms such behaviour (opposite-sign vorticity on both sides of each blade). These counter-rotating structures originate from different natures of flow characteristics at these

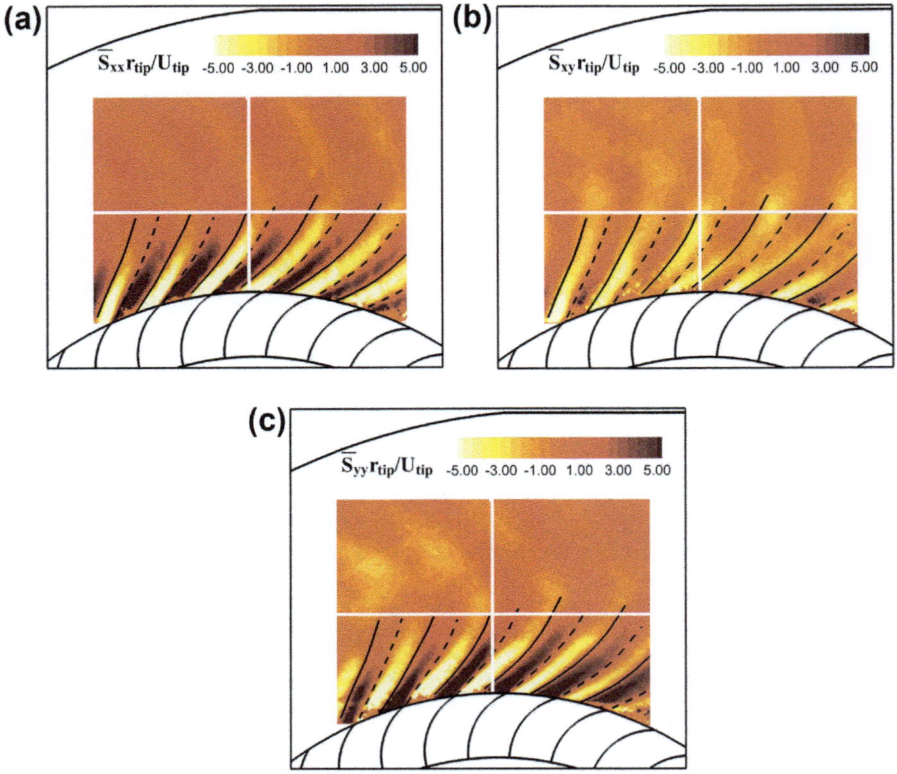

Figure 6.3 Ensemble average of normalized strain rate components: (a) the normal x-component; (b) the shear component in the x–y (measurement) plane and (c) the normal y-component.

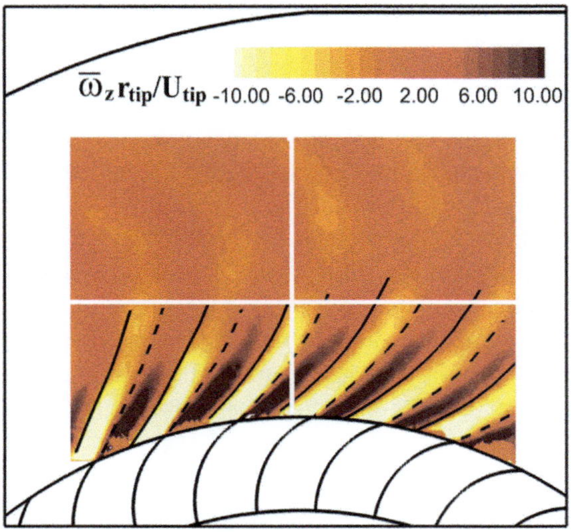

Figure 6.4 Ensemble average of normalized out-of-plane vorticity in the rotor exit region. The positive and negative values indicate counter-clockwise and clockwise flow structures (on average), respectively.

regions: the pressure-side flow is characterized by high-momentum jet stream while the suction-side structures are affected from separated low-momentum wake flow.

Figure 6.3(b) shows that maximum values of shear strain (regardless of its sign) occur in the interacting regions between each pair of jet and wake centrelines. These regions are strongly characterized by jet—wake interactions and cause considerable energy loss (dissipation) for the rotor exit flow. Blade congruency of flow is an important feature that reduces these losses and enhances turbomachine performance.

The cyclic pattern in the volute region is an evident feature in Figures 6.3 and 6.4. Interaction between rotor exit jet—wake structures with the volute flow generates cyclic or even spot-wise vortical structures with elevated level of deformation and vorticity. These interactions are also reported in axial turbomachines and are termed the consequence of wake—wake and wake—blade interactions. Their main effect is complication of rotor exit flow and generation of three-dimensional structures with elevated level of anisotropy. Two likely scenarios for such flow field are (a) increased energy loss and reduced turbomachine performance and (b) a serious challenge for accurate turbulence modelling (shall be discussed in Section 6.4).

6.3 Geometrical characteristics of tensorial-flow quantities

Knowledge about geometrical characteristics of flow quantities is required for further interpretation of flow interactions and evaluation of the consistency of turbulence models with the considered flow. The eigenvectors of tensorial-flow quantities such as strain rate and turbulence stress tensors are important contributors of turbulent-flow dynamics that are discussed in the following subsections.

6.3.1 *Interpretation of principal strain components*

Two-dimensional transformation of strain tensor into directions that maximize or minimize normal and shear components identifies geometrical structure of the complicated jet—wake flow. Considering ϕ as the angle of rotation from the original coordinate system into the second one, two-dimensional transformed matrix ($\mathbf{T}'$) of any two-dimensional tensor $\mathbf{T}$ can be calculated using the transformation matrix $\boldsymbol{\Psi}$ such that

$$\mathbf{T}' = \boldsymbol{\Psi}^{\mathbf{t}}\mathbf{T}\boldsymbol{\Psi} \tag{6.1}$$

where $\boldsymbol{\Psi}^{\mathbf{t}}$ is the transposition of tensor $\boldsymbol{\Psi}$ and

$$\mathbf{T}' = \begin{bmatrix} T'_{11} & T'_{12} \\ T'_{21} & T'_{22} \end{bmatrix}, \quad \mathbf{T} = \begin{bmatrix} T_{11} & T_{12} \\ T_{21} & T_{22} \end{bmatrix}, \quad \boldsymbol{\Psi} = \begin{bmatrix} \cos\phi & \sin\phi \\ -\sin\phi & \cos\phi \end{bmatrix} \tag{6.2}$$

Two principal coordinates are

1. *PN coordinate*[1] that corresponds to maximum normal component (11-direction) and minimum normal component (22-direction) with zero shear components.
2. *PS coordinate*[2] that corresponds to maximum shear components and aligns at 45° with the PN-coordinate system.

The principal directions and their respective principal components are in fact representatives of the eigenvectors and the eigenvalues of the tensor under consideration. Comparison of these eigenvalues (principal components) indicates the relative significance of their corresponding eigenvectors (principal directions) and consequently identifies the characterizing flow alignments in any complicated anisotropic flow field.

6.3.1.1 Geometrical characteristics of strain rate tensor

Principal in-plane components of ensemble-averaged strain tensor (from Eqn (6.1)) are shown in Figure 6.5 (strain components are normalized by U_{tip}/r_{tip}). The figure also presents alignments of maximum and minimum normal strains (PN coordinate system) and maximum shear strain (PS coordinate system). The maximum of all principal strain components is in the interacting regions between jets and wakes. According to this observation and smallness of three dimensionality in the jet–wake-interacting regions (shall be shown later), these regions experience strong planar straining.

Comparison of Figure 6.5(a)–(c) shows that jet–wake-interacting regions along the blades suction side (jet–wake suction side, JWSS) are aligned with maximum shear strain orientation. The difference between the alignments of maximum shear strain and the centreline of jet–wake-interacting regions along the rotor pressure side (jet–wake pressure side, JWPS) is about 90°. These regions correspond to minimum shear strain (with large magnitude) due to the 90° difference between alignments of minimum and maximum shear strain in the principal coordinate system. The opposite trends of principal shear components on two sides of each jet or wake structure are due to relative locations of jet–wake structures that generate angular deformations with opposite directions. If shear direction is excluded from analysis, the dominant shear flow occurs in the interacting regions between high- and low-momentum streams.

Figure 6.5(a) and (b) shows that the alignments of maximum normal strains are approximately perpendicular to jet and wake centrelines, and the alignments of minimum normal strains are roughly in line with jet and wake centrelines. Negative maximum normal strain and positive minimum normal strain are shown in white (enclosed regions) in Figure 6.5. According to Figure 6.5(b), the maximum normal strain is positive (extensive) in most regions of the rotor exit flow, while the minimum normal strain, Figure 6.5(c), is dominantly negative (contractive). Minimum negative values of flow straining along jets and wakes plus maximum positive values of flow straining perpendicular to jets and wakes show flattening of flow structures in planes

[1] Principal normal coordinate.
[2] Principal shear coordinate.

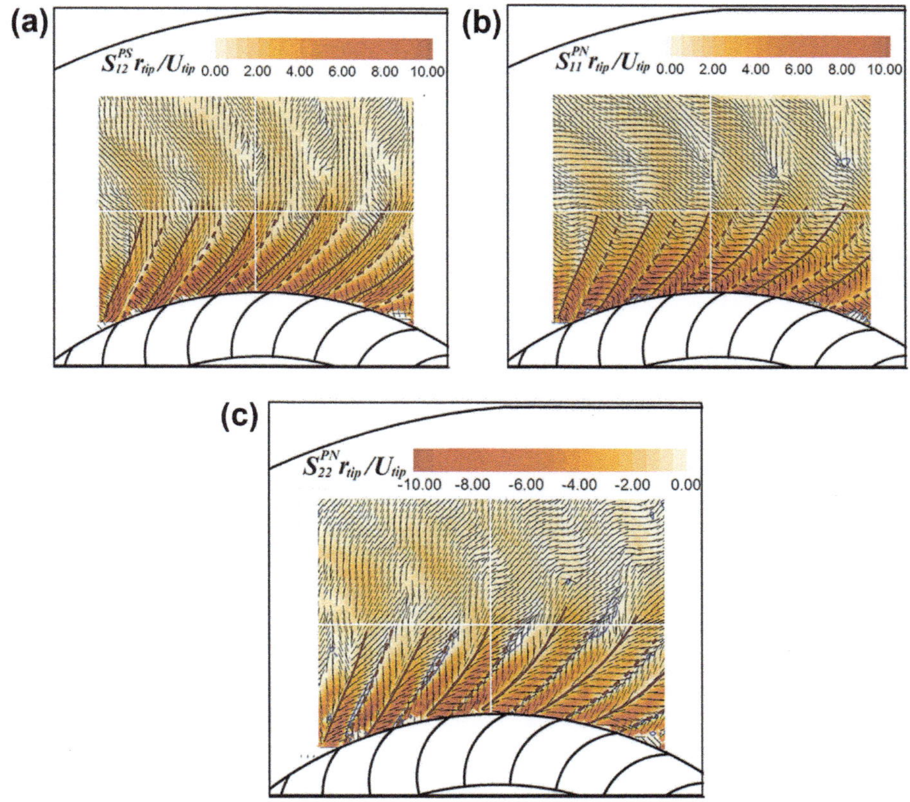

Figure 6.5 Variation of principal in-plane strain tensor components and orientation of principal coordinates. (a) Maximized shear strains and their corresponding coordinate orientation. (b) Alignments and values of maximum normal strain. Negative values are enclosed and shown in white. (c) Minimized normal strains and their corresponding coordinate orientation. Positive values are enclosed and shown in white.

normal to jet and wake centrelines. In other words, flow is decelerated as moves outward and structures are deformed in the normal direction. Since the flow field is dominantly planar, this phenomenon generates rod-like flow structures with one dominant eigenvalue. It is shown in the following section that one-component turbulence is also the dominant geometry for turbulence stress tensor in the rotor exit flow field.

6.3.1.2 Three dimensionality of jet–wake flow

Three-dimensionality factor (ζ) is defined as the ratio of out-of-plane longitudinal deformation to the maximum in-plane longitudinal deformations in the principal coordinate:

$$\zeta = \frac{|\langle S_{33}\rangle|}{\max\{|\langle S_{11}^{PN}\rangle|, |\langle S_{22}^{PN}\rangle|\}} \tag{6.3}$$

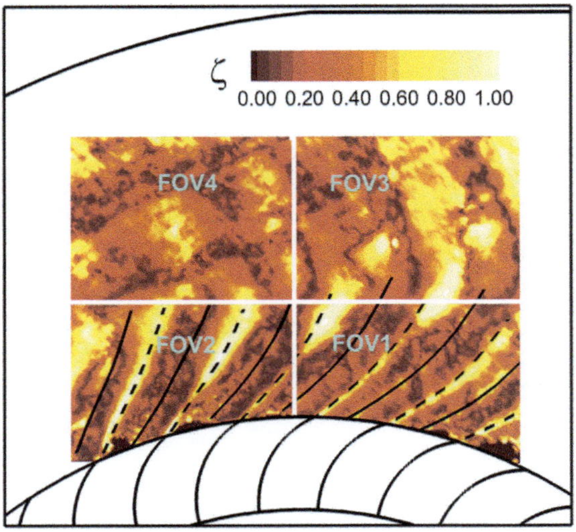

Figure 6.6 Three-dimensionality factor in the rotor exit region.

where $\langle \cdot \rangle$ denotes the ensemble-averaging operation. Three-dimensionality factor identifies the relative significance of out-of-plane to in-plane deformations.

Figure 6.6 presents the three-dimensionality factor in the rotor exit region of the present turbomachine. The flow at this region is strongly affected by interactions between jet—wake structures and volute flow. These interactions have stronger contributions to the low-momentum wake structures and create more significant out-of-plane deformations (elevation of three-dimensionality factor along wake centrelines in Figure 6.6).

The flow in the rotor exit of forward-curved centrifugal fan is not circumferentially symmetric even at design point. Consequently, all velocity-derived flow quantities such as strain tensor and three-dimensionality factor vary circumferentially. Figure 6.6 shows that the volute spacing rapidly extends in the right region of the measurement area (FOV1 and FOV3). This added extension further results in an asymmetric flow field.

6.3.2 Anisotropy of turbulence stress tensor

Presence of a wall in the flow domain causes anisotropic flow structures that are effective in complicating the geometrical properties and increase energy loss. Large Reynolds numbers intensify the extent of near-wall anisotropy due to stronger shear flows. The anisotropic flow structures in the complicated vortical flow of turbomachines are not limited to near-wall regions (rotor/stator blades, hub and casing). Examination of Reynolds stress tensor in the mixing tank shows that the impeller-induced flow generates elevated values of anisotropy in different near- and far-impeller locations. The geometry of turbulence stress tensor in the rotor exit region of squirrel-cage fan is expected to be complicated and directional. This is examined by measuring the extent of SGS stress tensor,

$$\tau_{ij} = \overline{u_i u_j} - \overline{u_i}\,\overline{u_j} \tag{6.4}$$

where the over-bar denotes filtering operation.

The Lumely triangle is a popular tool to examine the anisotropy behaviour of any tensor. It determines the anisotropy state of tensors in a coordinate system that consists of the second and the third invariants of anisotropy tensor. The target state coordinate inside the Lumley triangle determines the anisotropy state. The Lumley triangle is a nonlinear map and could lead to incorrect interpretation of turbulence states. A suitable alternative is the Barycentric map that determines the turbulence state in a linear triangular map.

The Barycentric map is a triangle with arbitrary vertices that correspond to three limiting states of turbulence, namely one-component, two-component axisymmetric and three-component states. The edges correspond to two-component, axisymmetric expansion and axisymmetric contraction limits. All feasible flow states locate inside the Barycentric triangle. Simpler flow fields generally result in more limited turbulence states (concentrated distribution of points in the invariant map), while complicated turbulent flows show a diverse range of turbulence states (wide distribution of points in the invariant map).

Figure 6.7 illustrates the Barycentric map for instantaneous SGS stress (hereafter called SGS stress). The points in this figure correspond to the state of turbulence in various locations of FOV1 (Figure 6.1) for one arbitrary snapshot. The turbulence considerably tends towards two-component or even one-component limiting states. The one-component state is the ultimate limit of an extensive structure that has a rod-like topology and the two-component limiting state corresponds to elliptical disk-like turbulence which is the ultimate limit of a contractive structure. Investigation of instantaneous SGS stress field in the complete jet–wake–volute area and various snapshots shows similar anisotropy trends.

The Barycentric map defines three limiting state weights, C_{1C}, C_{2C} and C_{3C}, that correspond to one-component, two-component axisymmetric and three-component states, respectively. In order to further clarify the tendency of SGS stress tensor in the Barycentric map to each of the three limiting states, probability density function (PDF) of the turbulence limiting state weights is presented in Figure 6.8 for various locations of FOV1. According to this graph, the maximum probability for the weight of one-component, two-component axisymmetric and three-component turbulence states are about 0.90, 0.04 and 0.01, respectively. Hence, the SGS stress has one dominant component along the eigenvector corresponding to the largest eigenvalue. The tendency to isotropic turbulence (three-component state) is very small for this complicated flow field. In other words, the rotor exit flow field is highly directional and anisotropic.

The anisotropy behaviour of SGS stress tensor in this rotor exit flow is quite different from that of classical turbulent case studies. Investigation of various turbulent flows such as isotropic turbulent flow, duct flow and boundary layer shows that the

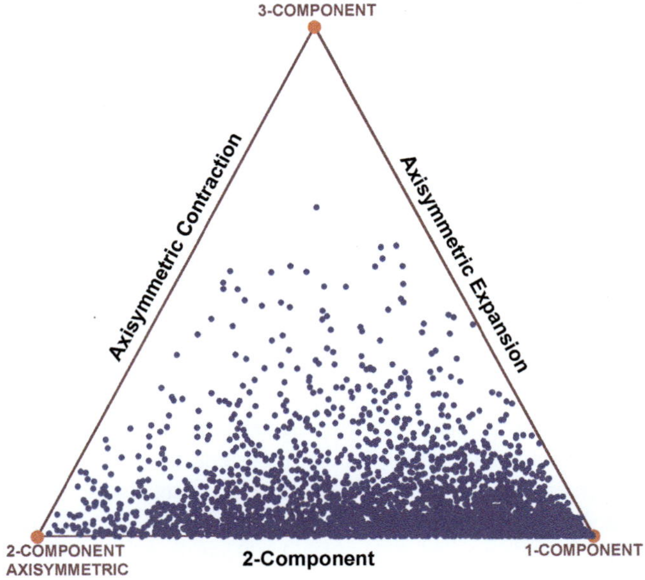

Figure 6.7 The turbulence state of instantaneous sub-grid scale stress in various locations of FOV1.

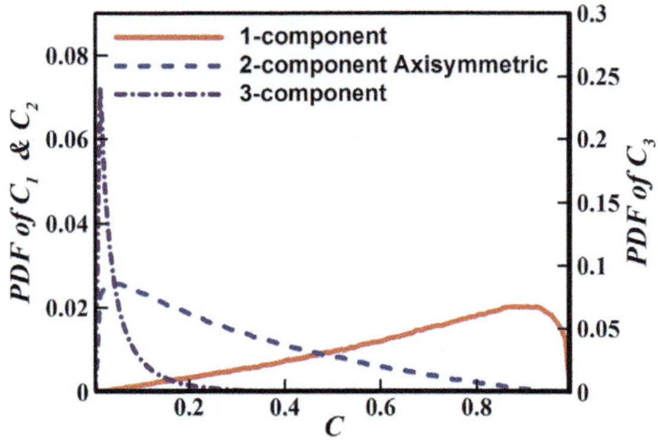

Figure 6.8 Probability density function of the limiting state weights for FOV1.

most probable state of anisotropy is axisymmetric extension for the fluid element deformation (strain rate tensor) and axisymmetric contraction for the SGS stress tensor.

6.4 Challenges in turbulence modelling

The separated flows, complicated geometry, large number of blades and small blade-to-blade distance are features in the forward-curved squirrel-cage fan that make an accurate prediction of turbulence stress difficult. The numerical grid should be fine enough to capture sufficient turbulent kinetic energy. Additionally, the turbulence model should be consistent with the physical nature of such complicated turbulent flow. Evaluation of various turbulence models in classical flow fields shows that anisotropy and complexity of flow reduce model performance. Turbulence models that are developed for limited states of turbulence are not appropriate for complicated flows that contain intensive temporal or spatial variations in turbulence states. Specific modifications to the limited-state-based models are required to make them applicable to such anisotropic flows, for example, using a damping function for the near-wall flows.

The modelling decomposition approach should be consistent with the characteristics of considered flow. Implementation of Reynolds-averaged Navier–Stokes approach for unsteady flow of turbomachines removes a significant part of directional information and is not appropriate for such temporally evolving flow field. The large eddy simulation (LES) approach decomposes flow scales into large (resolved) and small (unresolved or sub-grid) scales by filtering the velocity field. LES approach takes benefit from universality of small scales and models them using various SGS models, while directly calculating the resolved scales by solving the Navier–Stokes equations. It is an appropriate tool for simulating highly unsteady flows if at least the following two conditions are fulfilled:

1. The numerical grid and filter scale should be sufficiently fine to ensure a universal topology of stress tensor and also capture enough turbulent kinetic energy.
2. The SGS model should be consistent with the dynamics of eddies and turbulent energy-transfer mechanisms. This is a challenge for complicated flows such as turbomachinery ones.

Section 6.3 showed that strain and turbulence stress tensors in the rotor exit jet–wake flow are considerably anisotropic. Selection and implementation of an appropriate SGS model for such a complicated flow field require a priori evaluation of various models based on experimental data. It would even require modifying the available models to enhance the model accuracy in predicting SGS quantities.

6.4.1 Implementation of appropriate model coefficient

The eddy-viscosity (Smagorinsky) and similarity models are two SGS models that are frequently used in various flow simulations. The existing coefficients in these models are adjustable parameters that affect prediction accuracy. The value of Smagorinsky coefficient (C_{Smag}) is generally considered to be about 0.16, and a value about unity is selected for the similarity model coefficient. However, for a complicated flow field, using a unique coefficient is not efficient and it is more appropriate if the coefficient is tuned to local flow structures. The consistent coefficients for Smagorinsky and similarity models (C_{Smag}^{diss} and C_{Sim}^{diss}, respectively) can be calculated a priori through balancing the model-based SGS dissipation with the real calculated SGS dissipation:

$$2\left(\Delta C_{Smag}^{diss}\right)^2 \left\langle |\overline{S}| \overline{S_{ij}}\, \overline{S_{ij}} \right\rangle = -\left\langle (\overline{u_i u_j} - \overline{u_i}\, \overline{u_j}) \overline{S_{ij}} \right\rangle \tag{6.5}$$

$$C_{Sim}^{diss}\left\langle\left(\widetilde{\overline{u_i}\,\overline{u_j}} - \widetilde{\overline{u_i}}\,\widetilde{\overline{u_j}}\right)\overline{S_{ij}}\right\rangle = \left\langle\left(\overline{u_i u_j} - \overline{u_i}\,\overline{u_j}\right)\overline{S_{ij}}\right\rangle \tag{6.6}$$

where $|\overline{S}| = \sqrt{2\overline{S_{ij}}\,\overline{S_{ij}}}$ is the characteristic filtered strain rate (also known as strain magnitude). Since C_{Smag}^{diss} and C_{Sim}^{diss} are calculated on the basis of SGS dissipation, they are termed dissipation-based Smagorinsky or similarity coefficients, respectively. The filtered velocity and strain components in Eqns (6.5) and (6.6) are directly calculable by explicit implementation of a filter on SPIV data.

Table 6.1 summarizes the extracted dissipation-based model coefficients for two filter length scales, namely, $\Delta = 3\delta$ and $\Delta = 5\delta$ (in which $\delta = \sqrt{\delta_x\delta_y}$, where δ_x and δ_y are the spacing of regular SPIV grid in x- and y-directions, respectively). The coefficients are averaged over all points of each FOV. Table 6.1 shows that the average model coefficients for the jet–wake flow in the squirrel-cage fan are significantly smaller than their classical values (about 60–73% for the Smagorinsky model and 61–89% for the similarity model) and still generate sufficient energy flux among resolved and SGSs.

Figure 6.9 illustrates the distribution of dissipation-based model coefficients in four FOVs. A wide range of dissipation-based Smagorinsky and similarity coefficients can be observed in the contour plots of Figure 6.9. This wide range shows that Smagorinsky and similarity coefficients are very sensitive to flow structures. The Smagorinsky model is completely dissipative, so the left-hand side of Eqn (6.5) may never be negative. Consequently, the dissipation-based Smagorinsky coefficient is not defined wherever the real SGS dissipation (right-hand side of Eqn (6.5)) is negative. White regions in Figure 6.9(a) correspond to these undefined values. These are regions with dominant backscattering of turbulent energy (transfer of energy from SGS to resolved scales).

In regions with negative dissipation-based similarity coefficient, the predicted direction of turbulent energy flux by model (left-hand side of Eqn (6.6)) is opposite to

Table 6.1 Dissipation-based model coefficients for different filter length scales in four considered fields of view (FOVs)

		Filter length scale (Δ)	
Model coefficient	**FOV**	**3δ**	**5δ**
C_{Smag}^{diss}	FOV1	0.0501	0.0637
	FOV2	0.0484	0.0590
	FOV3	0.0531	0.0641
	FOV4	0.0434	0.0516
C_{Sim}^{diss}	FOV1	0.3877	0.2460
	FOV2	0.3537	0.1124
	FOV3	0.1964	0.3803
	FOV4	0.1655	0.2283

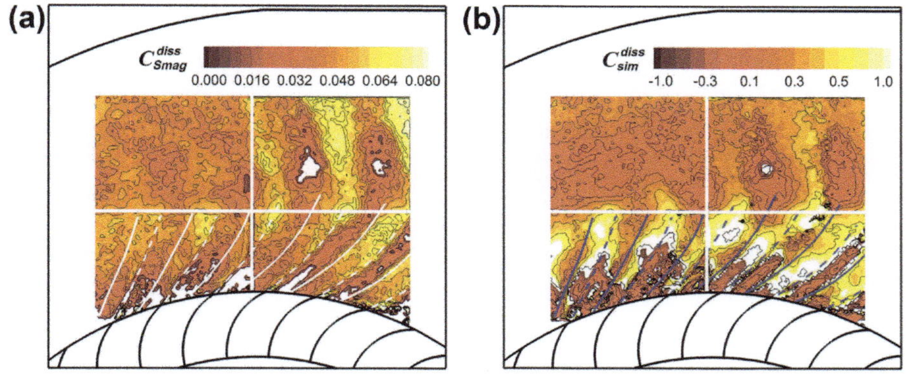

Figure 6.9 (a) The dissipation-based Smagorinsky coefficient; undefined values from Eqn (6.5) are shown in white. (b) The dissipation-based similarity coefficient; the coefficients larger than 1.0 and smaller than −1.0 are shown in white. Jet and wake trajectories are illustrated by solid and dashed curves, respectively.

that of real flow (right-hand side of Eqn (6.6)). In Figure 6.9(b), there are also a few regions with absolute magnitudes of similarity coefficient larger than unity (shown in white). In these narrow regions, the predicted SGS dissipation by similarity model (using classical similarity constant) is considerably lower than the real SGS dissipation. The similarity coefficient magnitude should be increased to compensate this discrepancy.

Increasing the density of flow samples with energy backscatter decreases applicability of fully dissipative models such as Smagorinsky model. The PDF of normalized SGS dissipation is illustrated in Figure 6.10 for instantaneous flow samples for different regions of the rotor exit flow: jet and wake centrelines, jet–wake-interacting region along the blades pressure side (JWPS) and jet–wake-interacting region along the blades suction side (JWSS).

According to the PDFs of Figure 6.10, about 40% of instantaneous flow samples contain energy backscattering. Additionally, the distribution of PDFs is fairly symmetric and many flow samples with intensive backward energy transfer are present in all flow regions. Eddy-viscosity models are therefore not appropriate candidates for SGS energy flux prediction in the rotor exit region of squirrel-cage fans.

6.4.2 A priori evaluation of model accuracy

A common approach to evaluate the accuracy of an SGS model is to calculate the correlation coefficient between real and modelled SGS dissipation (Π^{real} and Π^{mod}, respectively):

$$\rho\left(\Pi^{mod},\ \Pi^{real}\right) = \frac{\left\langle \Pi^{mod}\ \Pi^{real}\right\rangle - \left\langle \Pi^{mod}\right\rangle\left\langle \Pi^{real}\right\rangle}{\left[\left(\left\langle\left(\Pi^{mod}\right)^2\right\rangle - \left\langle \Pi^{mod}\right\rangle^2\right)\left(\left\langle\left(\Pi^{real}\right)^2\right\rangle - \left\langle \Pi^{real}\right\rangle^2\right)\right]^{1/2}} \tag{6.7}$$

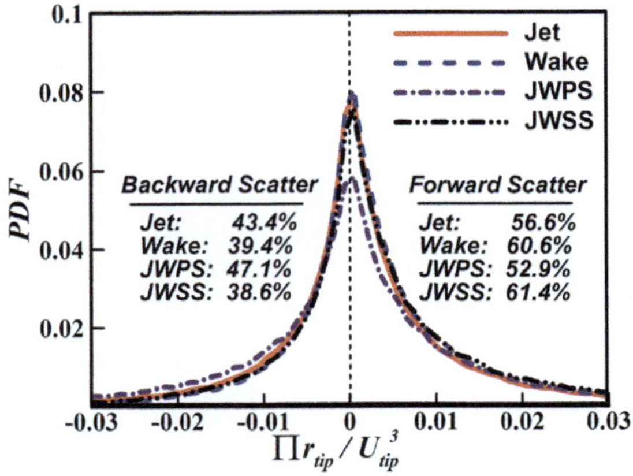

Figure 6.10 Probability density function of instantaneous sub-grid scale (SGS) dissipation for various temporal and spatial flow data. The SGS dissipation is normalized by U_{tip}^3/r_{tip}.

A correlation coefficient equal to 1 is an exact estimation where multiplication of the modelled quantity by a scalar coefficient would give the original quantity. On the other hand, models with correlation coefficients close to zero are almost uncorrelated with the real quantities and should not be implemented in turbulence simulations.

Table 6.2 presents the correlation coefficient of SGS dissipation (Eqn (6.7)) based on different filter length scales. Correlation coefficients in each FOV are averaged over all snapshots and entire velocity points of that FOV. It is clear that the similarity model is superior to the Smagorinsky model, although it still has considerable deviation from accurate prediction (correlation coefficient of unity).

Table 6.2 Correlation coefficient between real and modelled SGS dissipations for two filter length scales in four fields of view (FOVs)

Correlation coefficient	FOV	Filter length scale (Δ)	
		3δ	5δ
$\rho(\Pi^{Smag},\Pi^{real})$	FOV1	0.2206	0.1806
	FOV2	0.3045	0.2355
	FOV3	0.2362	0.2329
	FOV4	0.3871	0.3847
$\rho(\Pi^{Sim},\Pi^{real})$	FOV1	0.4932	0.2298
	FOV2	0.5765	0.3758
	FOV3	0.6709	0.4079
	FOV4	0.7002	0.5117

Comparison of SGS dissipation correlation coefficients for two filter length scales in Table 6.2 shows that the correlation reduces by increasing the filter scale. When a larger filter length scale is employed, modelled quantities are calculated based on a wider flow region and consequently larger and more anisotropic flow structures. These larger structures are less universal as compared with fine-scale structures and result in more inaccurate models with smaller correlation coefficients.

The SGS models can also be evaluated by examining the joint PDF of real calculated SGS stress/dissipation versus the modelled SGS stress/dissipation. Figure 6.11 illustrates the joint PDF of real SGS stress components (normal and shear components) with respect to the Smagorinsky stress components. A linear functionality in such a joint PDF map means that the model is exact, and closer probability distribution to a linear relation indicates that the model is more capable in predicting correct SGS stress. The dashed linear line in Figure 6.11 corresponds to exact modelling of SGS stress. Wide distribution of probability around this line and absence of a specific alignment in the joint PDFs of Figure 6.11 indicate that the Smagorinsky model is not able to predict SGS stress accurately. Another remark about Figure 6.11 is the existence of negative and positive values for normal components of Smagorinsky-modelled SGS stress, while the normal components of SGS stress as calculated from the definition are positive everywhere. The Smagorinsky model is not able to correctly predict the SGS stress sign for normal components which certainly affects the prediction of turbulent energy flux.

The joint PDF of real SGS stress components (normal and shear components) versus the similarity-modelled SGS stress components is presented in Figure 6.12 for FOV1. The PDFs are drawn for two different similarity constants: $C_{Sim} = 1.0$ which is commonly used for various flow fields and $C_{Sim} = 0.2$. Unlike the

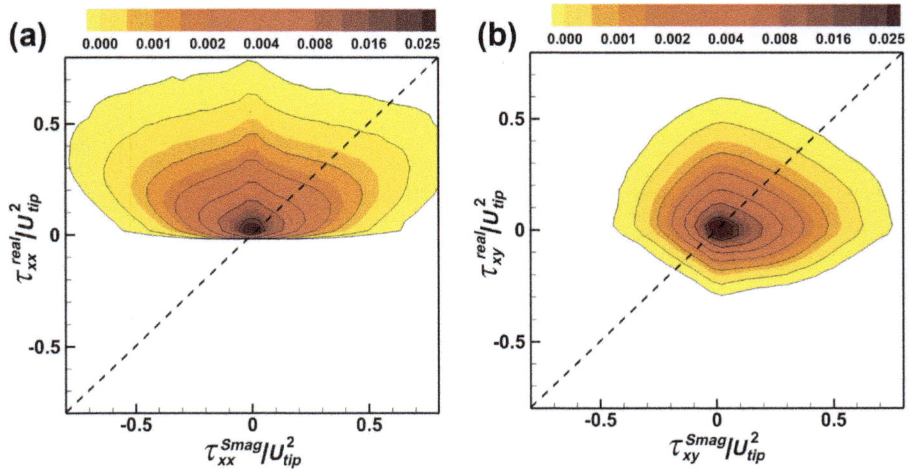

Figure 6.11 Joint probability density function (PDF) of real sub-grid scale (SGS) stress versus Smagorinsky-modelled SGS stress for FOV1: (a) normal xx-component, (b) shear xy-component. The dashed lines indicate the joint PDFs for an exact model. The SGS stress is normalized by U_{tip}^2.

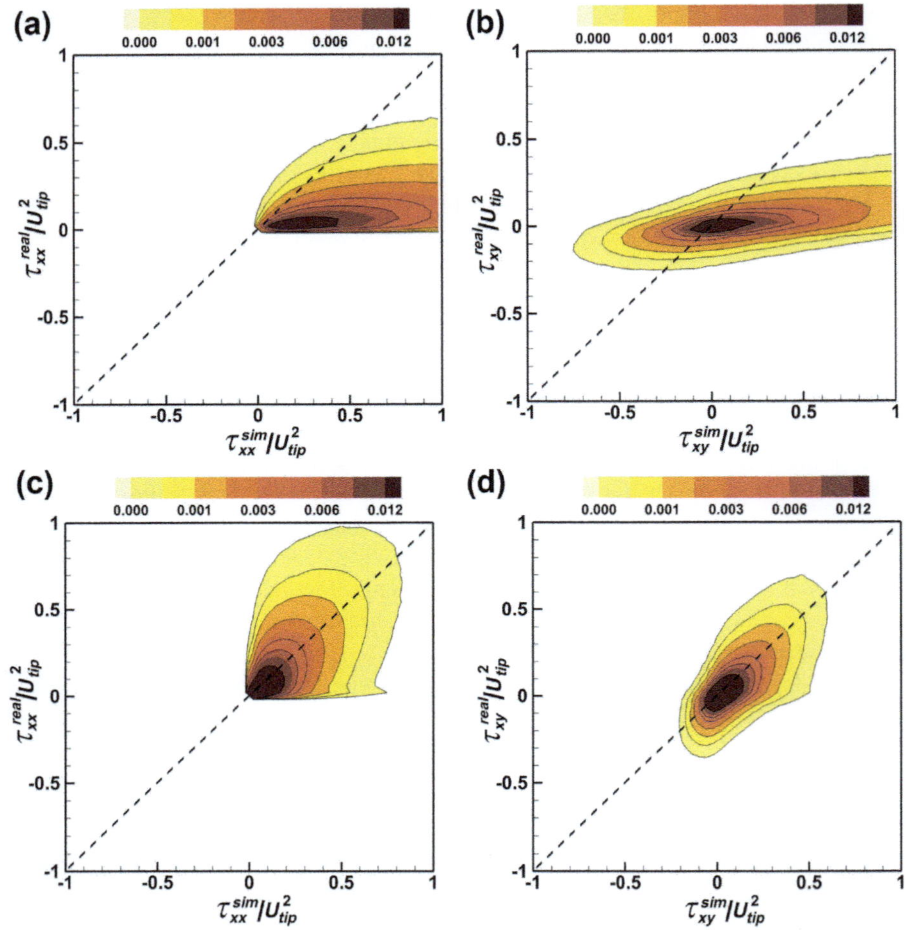

Figure 6.12 Joint probability density function (PDF) of real versus similarity-modelled sub-grid scale (SGS) stress components: (a) xx-component, $C_{sim} = 1.0$; (b) xy-component, $C_{sim} = 1.0$; (c) xx-component, $C_{sim} = 0.2$; and (d) xy-component, $C_{sim} = 0.2$. The dashed lines indicate the joint PDFs for an exact model. The SGS stress is normalized by U_{tip}^2.

Smagorinsky model, there are stronger alignments in the joint PDFs of real versus similarity-modelled SGS stress components (joint PDF of yy-component of SGS stress was similar to the xx-component and is not shown here). In the case of $C_{Sim} = 1.0$, this alignment is not consistent with the expected alignment for exact prediction of SGS stress (the dashed lines in the presented PDFs). The optimum alignment is for $C_{Sim} = 0.2$, where the distribution is considerably closer to an exact prediction (Figure 6.12).

Another remark about Figure 6.12 is the ability of similarity model in correct prediction of the sign of normal SGS stress components, while the Smagorinsky model is not able to do this prediction.

Investigation of joint PDFs of real SGS dissipation versus similarity-modelled SGS dissipation shows similar trends (not presented here). The prediction of SGS dissipation by similarity model was also optimum in the rotor exit region for $C_{sim} = 0.2$.

6.4.3 Addressing the turbulence-modelling development

According to the discussed challenges in the turbulence modelling of jet—wake flow, accurate prediction of turbulence stress tensor requires model enhancement consistent with flow dynamics. Nonlinear and nonlocal energy-transfer mechanisms are responsible for altering turbulent energy cascade and have significant contribution in SGS model accuracy. Further understanding of such transfer mechanisms in highly anisotropic flows is required to develop more precise SGS models.

A possible approach would be to determine the contributing parameters in the stress tensor and using statistical tools to express it as a function of resolved quantities. This enhancement is not the sufficient condition and resolution of numerical grid should be appropriately fine. This requires an enormous computational resource which is limited at present. In fact, the advantages of any SGS model evaluation for complicated turbomachinery applications will be revealed in the future by further progress of computational resources.

6.5 Final remarks

The SPIV shows that the rotor exit flow is characterized by significant jet—wake—volute interactions. Mixing of high-momentum jets and retarding low-momentum wakes generate shear layers along the suction and pressure side of blades. Counter-rotating vortices and elevated angular deformations are the main characteristics in the opposite sides of each blade. Deviation of fan-operating condition from its design point intensifies jet—wake interactions, increases energy loss and decreases the machine performance.

The geometrical characteristics of flow structures show that planar or even rod-like structures are the dominant topology of flow deformations. However, these topologies are not necessarily the case in other sections of the rotor exit region. For instance, the recirculation zone in near-inlet region of the rotor exit flow experiences highly three-dimensional flow. The common point in every section of squirrel-cage fan is that the flow field is quite directional, even at design point.

Examination of anisotropy state by the Barycentric approach confirms such behaviour for turbulence stress tensor. High level of anisotropy is a challenge for turbulence modelling and simulation of flow field in every turbomachinery application. Presence of wall-induced fine-scale flow structures downstream of the rotor blades is another prominent challenge for simulation of turbomachinery flow. A priori evaluation of classical turbulence models on the basis of extracted SPIV data indicates that these models are not able to accurately predict turbulence stress tensor. The main reason is the limiting assumptions of such models that are not valid in the complicated flow of squirrel-cage fans. Modification of turbulence-model constants is helpful for enhancement of model performance. However, more accurate prediction of turbulence structures in such complicated geometries requires further development of turbulence models as well as computational resources.

References

Akbari, G., & Montazerin, N. (2013). A-priori study of subgrid-scale models for the flow field in the rotor exit region of a centrifugal turbomachine. *International Journal of Heat and Mass Transfer, 66*, 423–439.

Akbari, G., & Montazerin, N. (2013). On the role of anisotropic turbomachinery flow structures in inter-scale turbulence energy flux as deduced from SPIV measurements. *Journal of Turbulence, 14*, 44–70.

Akbari, G., Montazerin, N., & Akbarizadeh, M. (2011). Stereoscopic particle image velocimetry of the flow field in the rotor exit region of a forward-blade centrifugal turbomachine. *Proceedings of the Institution of Mechanical Engineers, Part A: Journal of Power and Energy, 226*, 163–181.

Banerjee, S., Krahl, R., Durst, F., & Zenger, C. (2007). Presentation of anisotropy properties of turbulence, invariants versus eigenvalue approaches. *Journal of Turbulence, 8*, 1–27.

Chamecki, M., Meneveau, C., & Parlange, M. B. (2007). The local structure of atmospheric turbulence and its effect on the Smagorinsky model for large eddy simulation. *Journal of the Atmospheric Sciences, 64*, 1941–1958.

Chow, Y. C., Uzol, O., & Katz, J. (2002). Flow nonuniformities and turbulent 'hot spots' due to wake-blade and wake-wake interactions in a multi-stage turbomachine. *Journal of Turbomachinery, 124*, 553–563.

Chow, Y. C., Uzol, O., Katz, J., & Meneveau, C. (2003). Experimental study of the structure of a rotor wake in a complex turbomachinery flow. In *4th ASME-JSME joint fluids Engineering Conference, Honolulu, Hawaii, USA*.

Derksen, J. J., Doelman, M. S., & Van Den Akker, H. E. A. (1999). Three-dimensional LDA measurements in the impeller region of a turbulently stirred tank. *Experiments in Fluids, 27*, 522–532.

Escudié, R., & Liné, A. (2006). Analysis of turbulence anisotropy in a mixing tank. *Chemical Engineering Science, 61*, 2771–2779.

Galletti, C., Brunazzi, E., Pintus, S., Paglianti, A., & Yianneskis, M. (2004). A study of Reynolds stresses, triple products and turbulence states in a radially stirred tank with 3-D laser anemometry. *Chemical Engineering Research and Design, 82*, 1214–1228.

Higgins, C. W., Parlange, M. B., & Meneveau, C. (2003). Alignment trends of velocity gradients and subgrid-scale fluxes in the turbulent atmospheric boundary layer. *Boundary-Layer Meteorology, 109*, 59–83.

Jaberi, F. A., & Colucci, P. J. (2003). Large eddy simulation of heat and mass transport in turbulent flows. Part 1: velocity field. *International Journal of Heat and Mass Transfer, 46*, 1811–1825.

Jovičić, N., Breuer, M., & Jovanović, J. (2006). Anisotropy-invariant mapping of turbulence in a flow past an unswept airfoil at high angle of attack. *Journal of Fluids Engineering, 128*, 559–567.

Liu, S., Meneveau, C., & Katz, J. (1994). On the properties of similarity subgrid-scale models as deduced from measurements in a turbulent jet. *Journal of Fluid Mechanics, 275*, 83–119.

Lumley, J. L., & Newman, G. (1977). The return to isotropy of homogeneous turbulence. *Journal of Fluid Mechanics, 82*, 161–178.

Lund, T., & Rogers, M. (1994). An improved measure of strain state probability in turbulent flows. *Physics of Fluids, 6*, 1838–1847.

Meneveau, C., & Katz, J. (2000). Scale-invariance and turbulence models for large-eddy simulation. *Annual Review of Fluid Mechanics, 32*, 1–32.

Montazerin, N., Damangir, A., & Mirian, S. (1998). A new concept for squirrel-cage fan inlet. *Proceedings of the Institution of Mechanical Engineers Part A Journal of Power and Energy, 212*, 343–349.

Morrison, J. F., Jiang, W., McKeon, B. J., & Smits, A. J. (2002). Reynolds-number dependence of streamwise velocity fluctuations in turbulent pipe flow. *The American Institute of Aeronautics and Astronautics*, 0574.

Porté-Agel, F., Parlange, M. B., Meneveau, C., & Eichinger, W. E. (2001). A priori field study of the subgrid-scale heat fluxes and dissipation in the atmospheric surface layer. *Journal of the Atmospheric Sciences, 58*, 2673–2698.

Simonsen, A. J., & Krogstad, P. A. (2005). Turbulent stress invariant analysis: clarification of existing terminology. *Physics of Fluids, 17*, 088103.

Tao, B., Katz, J., & Meneveau, C. (2002). Statistical geometry of subgrid-scale stresses determined from holographic particle image velocimetry measurements. *Journal of Fluid Mechanics, 457*, 35–78.

Vreman, B., Geurts, B., & Kuerten, H. (1994). On the formulation of the dynamic mixed subgrid-scale model. *Physics of Fluids, 6*, 4057–4059.

Index

Note: Page numbers with "f" denote figures.